THE BOOK OF FIRE

THE BOOK OF FIRE

William H. Cottrell Jr.

Mountain Press Publishing Company
in cooperation with the National Park Foundation
1989

This Book is Dedicated to
Mom and Dad
whose parenting permitted me to be
more curious than cautious about life's mysteries.

The Murphy-Phoenix Company
generously granted the financial support
which made possible the production of this book.
In publishing, a colored picture is worth a thousand words.

Illustration, design, and production by Dave Comstock

Library of Congress Cataloging-in-Publication Data

Cottrell, William H.
The book of fire.
1. Forest fires. 2. Fire. I. Title.
SD421.C714 1989 634.9'618 89-13052
ISBN 0-87842-255-2

Published by Mountain Press
P. O. Box 2399, Missoula, Montana 59806

CONTENTS

PREFACE

> "If it can't be expressed in figures, it is not a science; it is opinion."
> —Robert A. Heinlein

> "If it can't be explained with pictures, it can't be explained to me."
> —William H. Cottrell Jr., M.D.

Since Prehistory, man has stared into fire and wondered about flames, glowing coals, white ash and charcoal. Clear and understandable answers to obvious questions about the burning process were unavailable until now.

This book addresses the tantalizing "why" questions that fire proposes to the curious student of nature. Though deceptively simple in appearance, flame requires an understanding of the elemental processes by which fire transforms a candle, campfire, or forest into gases, charcoal, and ash. **Visualization** of the unseen universe of atoms and molecules immensely facilitates this understanding.

The book provides the reader with **impressions** from scientifically accurate illustrations and provides enough text and definitions to satisfy a wide variety of intellectual needs and time requirements.

Section I introduces the fundamental concepts of molecular physics, chemistry, and biology necessary to **understand** the obscure path of energy flow from its origin in the sun to its capture and storage by living cells and subsequent release by the chemical reaction known as **fire**.

Section II visually explores the "anatomy and physiology" of flame, and details the combustion process.

Section III applies concepts and vocabulary acquired from I and II to the real world of fire in forests.

Section IV explores the aftermath of forest fire and the physical clues that remain to reveal much about the type of forest fuel and the characteristic manner by which it burned.

ACKNOWLEDGMENTS

The following individuals, alphabetically listed, contributed their valuable time and intellect to the production of this manuscript:

Don Despain, Park Biologist in the Research Division, Yellowstone National Park. Thank you for going out into the firescapes and for manuscript critique.

Brian Jenkins, Ph.D., Associate Professor, Agricultural Engineering Department, University of California at Davis. Thank you for knowledge about cellulose combustion.

Ian Kennedy, Ph.D., Associate Professor, Chemical Engineering Department, University of California at Davis. Thank you for the long hours of questions and answers about flame and gas combustion.

George B. Robinson, Chief of Interpretation, Yellowstone National Park. Thank you for recognizing the need for general knowledge about fire science and telling me about it—for all your efforts in fund-raising, for your manuscript critique, and for your home phone number.

Richard Rothermel, Project Leader at the Intermountain Fire Research Lab, Missoula, Montana. Thank you for your immensely invaluable red-pencil criticisms and positive support toward scientific accuracy and currency.

Donald Ward, Ph.D., Project Leader at the Intermountain Fire Research Lab, Missoula, Montana. Thank you for your critical review of the manuscript.

Henry Shovik, Ph.D., Soils Scientist, Yellowstone National Park/Gallatin National Forest. Thanks for your positive support and knowledge about soils and forest regeneration after fire.

Mary Jane Sligar, M.S., "The non-scientist point of view." Thank you for clear-thinking observations and editing.

Lee, wherever you are.

I. ATOMS, MOLECULES, AND CHEMICAL REACTIONS

"The earth was created a living creature endowed with a soul and intelligence by the providence of God." —Plato

Fire describes the very rapid release of energy stored in food and fuel. The release of that stored energy involves complex chemical operations.

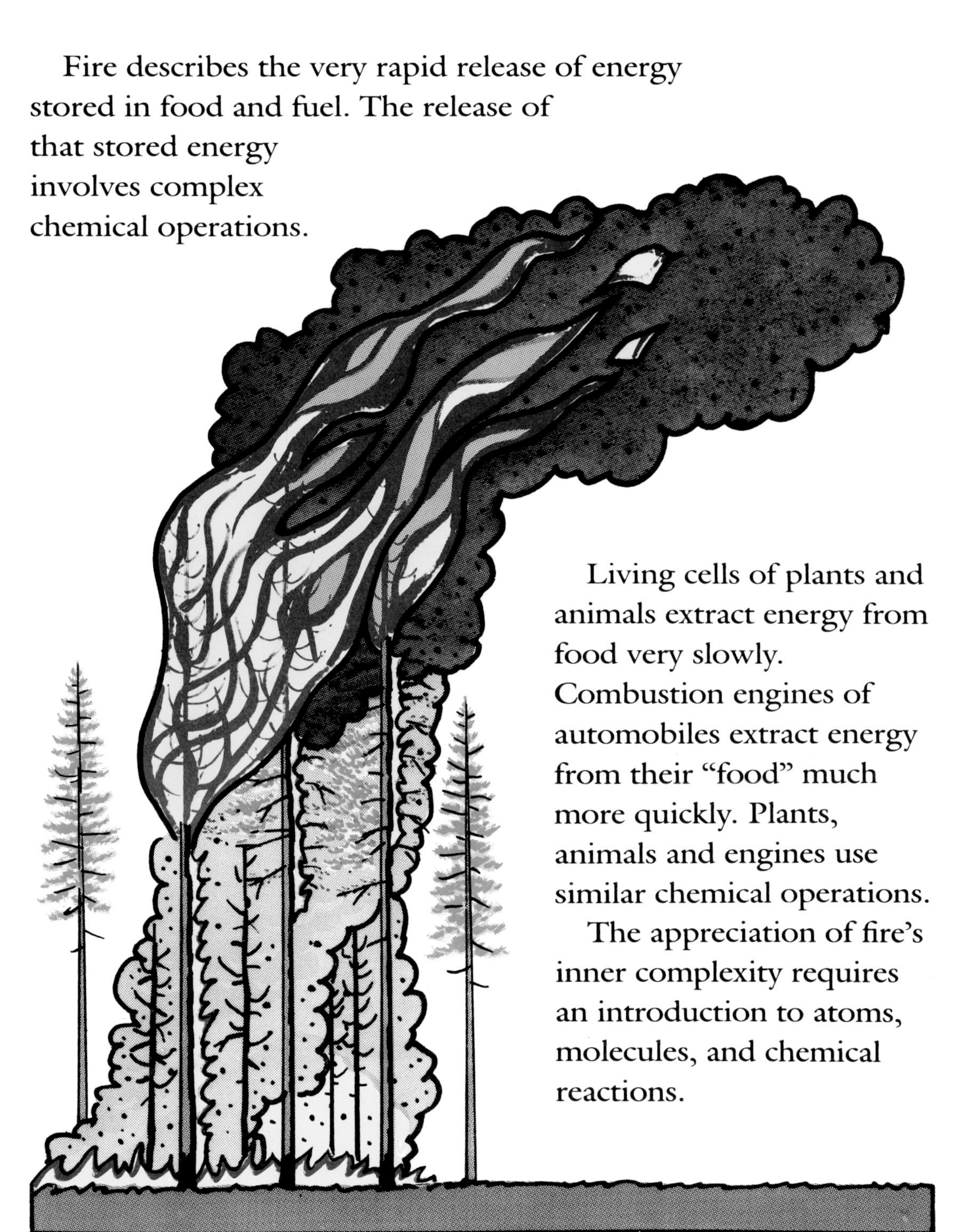

Living cells of plants and animals extract energy from food very slowly. Combustion engines of automobiles extract energy from their "food" much more quickly. Plants, animals and engines use similar chemical operations.

The appreciation of fire's inner complexity requires an introduction to atoms, molecules, and chemical reactions.

Earth, water, and air represent the three physical states of **matter**: **solid**, **liquid**, and **gas**. Nature uses tiny bits of matter called **atoms** and **molecules** to make mountains, forests, rivers, and air.

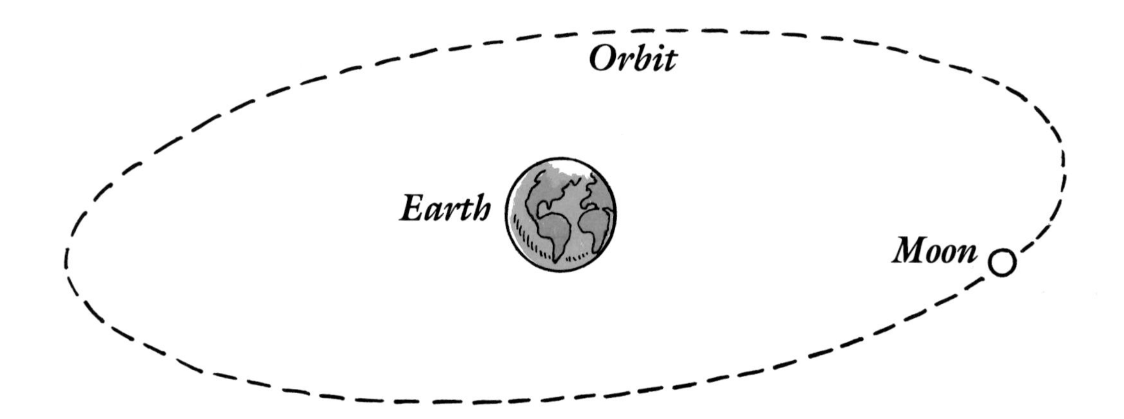

The earth and its orbiting moon present a visual model of a generic atom. An electron (moon) orbits the nucleus (earth) at a great distance and in relatively slow motion.

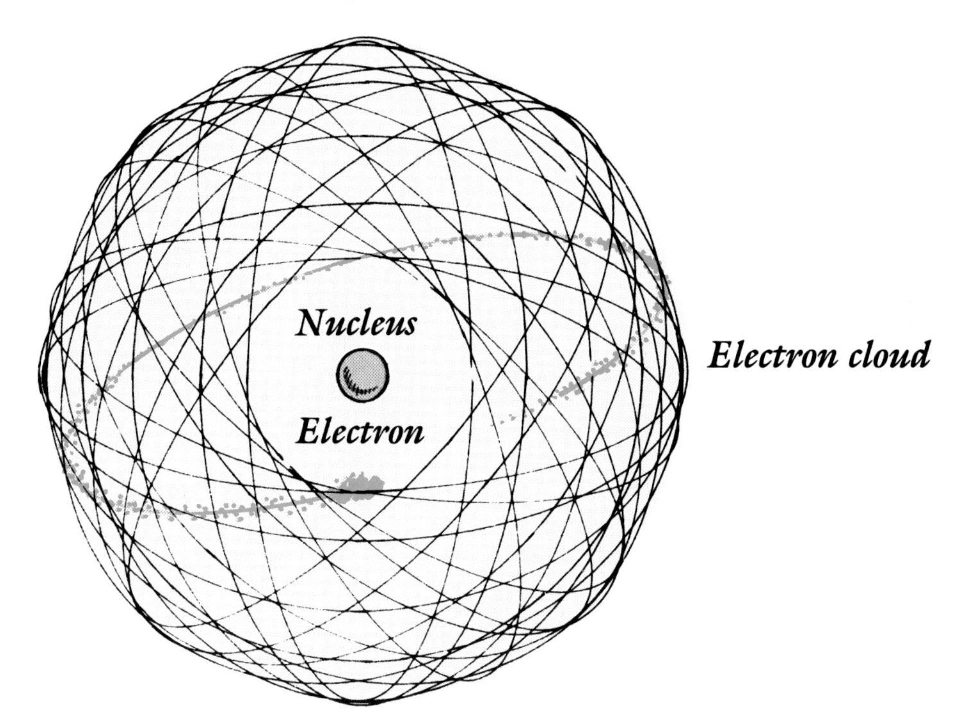

In reality, the orbiting electrons whiz about the nucleus so astonishingly fast and change orbits so frequently that the electrons appear to be everywhere at once and form a cloud-like shell enclosing the nucleus.

Atoms combine with other atoms to form **molecules** by merging their shells and sharing electrons. Even the largest molecules are exceedingly small, measuring one hundred millionth of an inch in diameter and weighing one million trillionths of a pound. A teaspoon (one cubic inch) of air contains four hundred thousand trillion molecules of various gases, mostly nitrogen and oxygen.

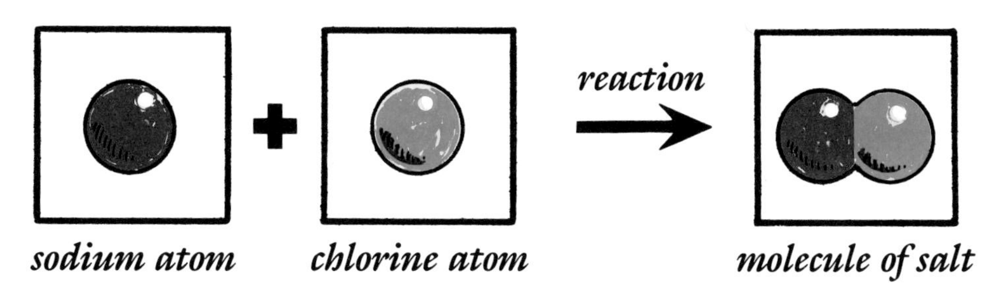

sodium atom *chlorine atom* *molecule of salt*

One **atom** of sodium and one **atom** of chlorine combine to form one **molecule** of sodium chloride (table salt).

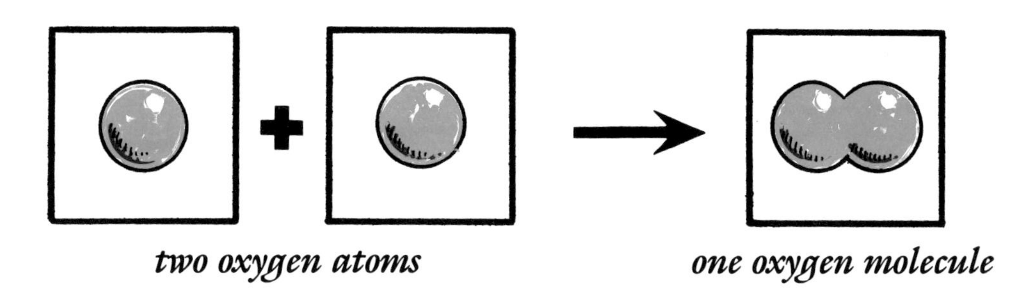

two oxygen atoms *one oxygen molecule*

Oxygen from the atmosphere usually consists of two atoms combined to form one molecule of oxygen.

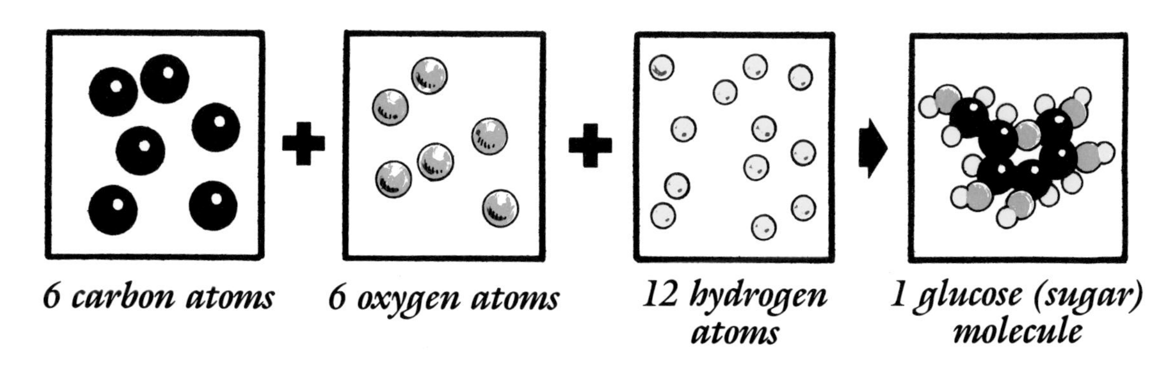

6 carbon atoms *6 oxygen atoms* *12 hydrogen atoms* *1 glucose (sugar) molecule*

Six atoms of carbon, twelve atoms of hydrogen, and six atoms of oxygen combine to form one molecule of sugar.

Unless "frozen" into a solid state of matter, molecules bounce about, randomly colliding with other molecules (five billion times per second for molecules in air). Heat energy stimulates the molecules into continuous activity. Without heat, molecular motion ceases.

This single molecule of oxygen in a container has been cooled to absolute zero degrees Kelvin (-273°C or -460°F). No heat energy transfers to the molecule to give it motion.

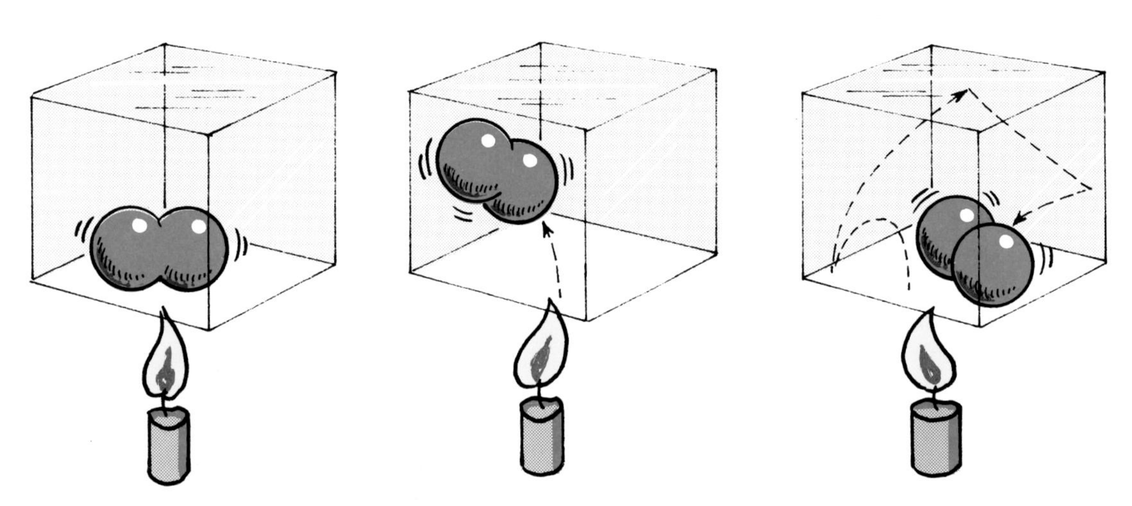

If heat is applied to the container, heat energy transfers to the molecule, which starts to vibrate, then bounce about the container. The molecular motion increases with continued heating.

If the container held six molecules, they would collide with each other. The number of collisions per second (rate) would be determined by the amount of heat absorbed by each molecule.

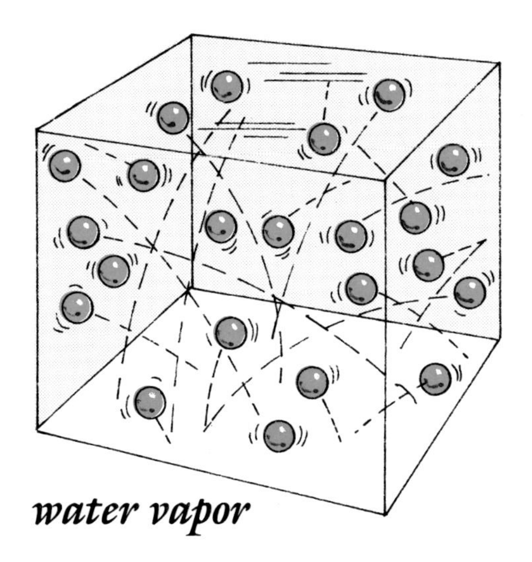

Water molecules provide an excellent analogy for molecular behavior.

This container demonstrates water molecules bouncing randomly in their gaseous state of **water vapor**. These molecules possess a lot of energy in their vapor state.

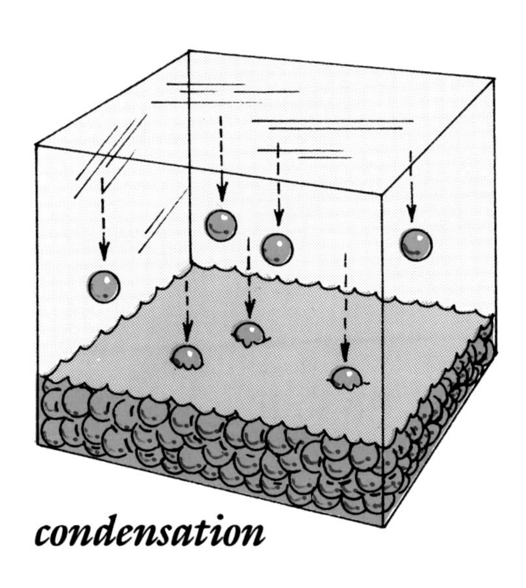

When gas molecules cool by giving off some of their heat energy, they slow down and transform into water molecules by a process called **condensation**. The cooler and less energetic molecules fall to collect as a liquid, where they still move about, but not nearly as much as when in the gas/vapor state.

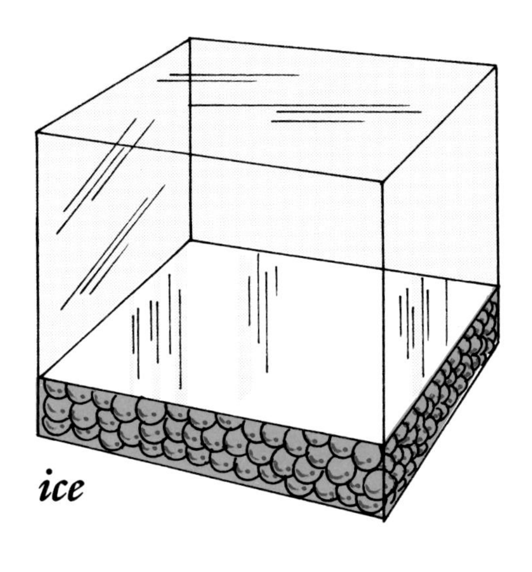

With further heat loss (cooling), the molecules slow even more and "freeze" into the solid form of water molecules called **ice**. The molecules, now trapped in a sold crystal network, move very little.

If the frozen-solid water molecules are now warmed, each molecule will start to vibrate as it absorbs heat energy.

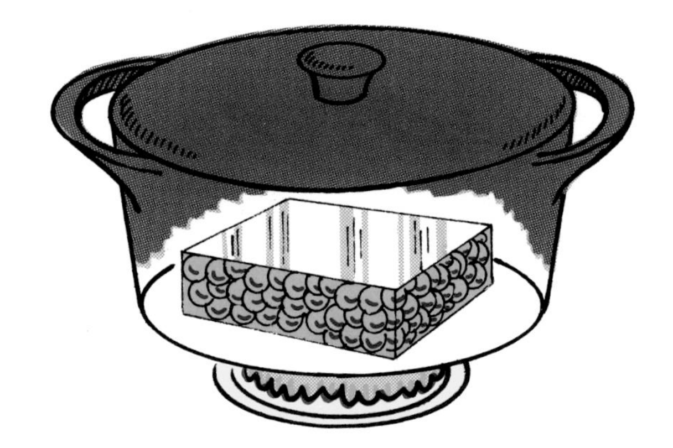

The molecules, when sufficiently heat-energized, transform from solid crystal to liquid water.

Continued molecular heating transforms the water molecules into water vapor, the gaseous form of water. The rapidly moving water molecules exert enough pressure to lift the pot's lid and escape into the air.

A **chemical reaction** describes what changes occur when different types of atoms or molecules interact.

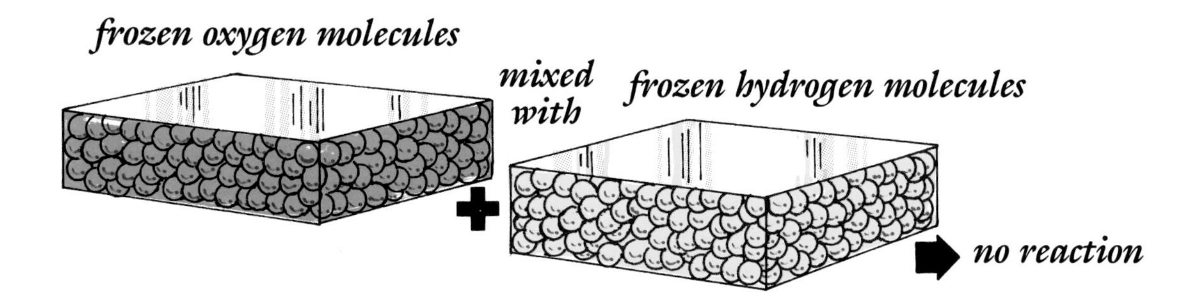

Chemical reactions need energy from an outside source to stir the molecules so they can participate in the reaction. Frozen molecules don't interact, so there is no reaction.

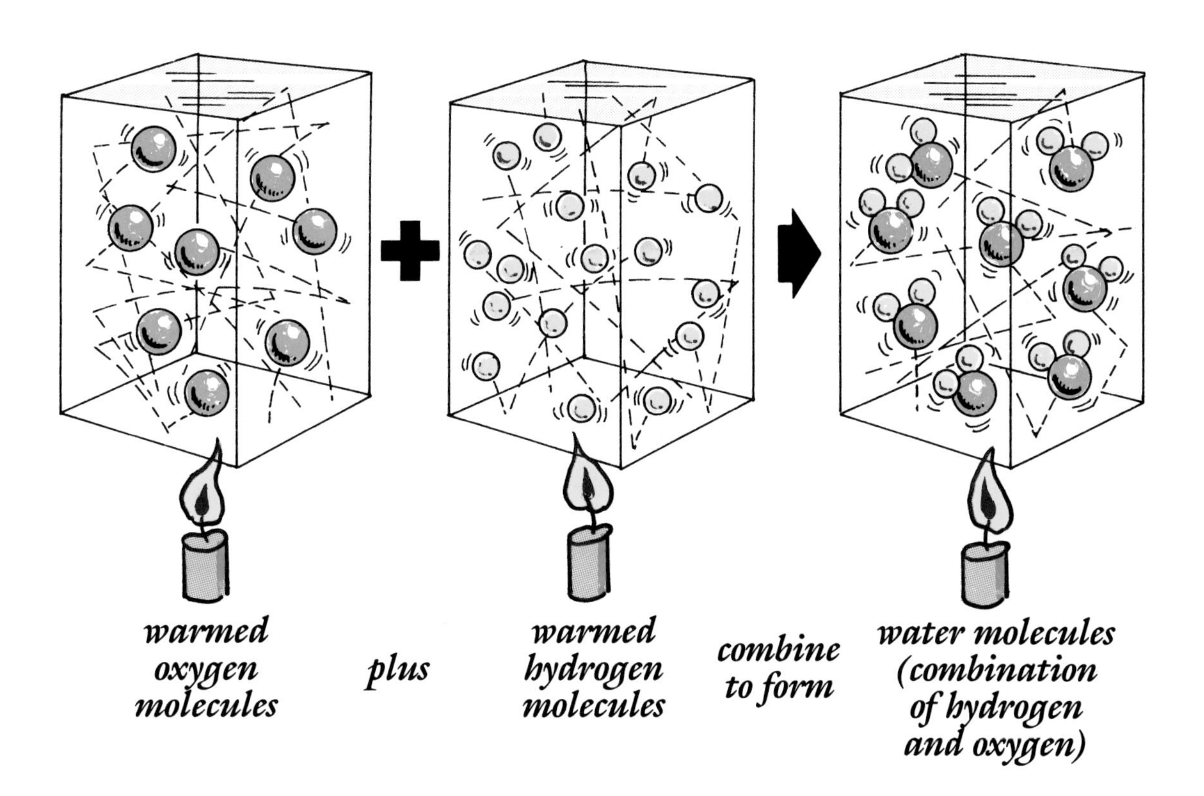

Warm molecules vibrate and bounce about freely and interact readily in chemical reactions if appropriate conditions exist. Heat governs the **rate of reaction**. Oxygen and hydrogen molecules **react** by sharing and exchanging electrons in their shells to form water molecules.

SUN ENERGY

The sun ultimately provides the energy required to keep molecules and atoms interacting in chemical reactions.

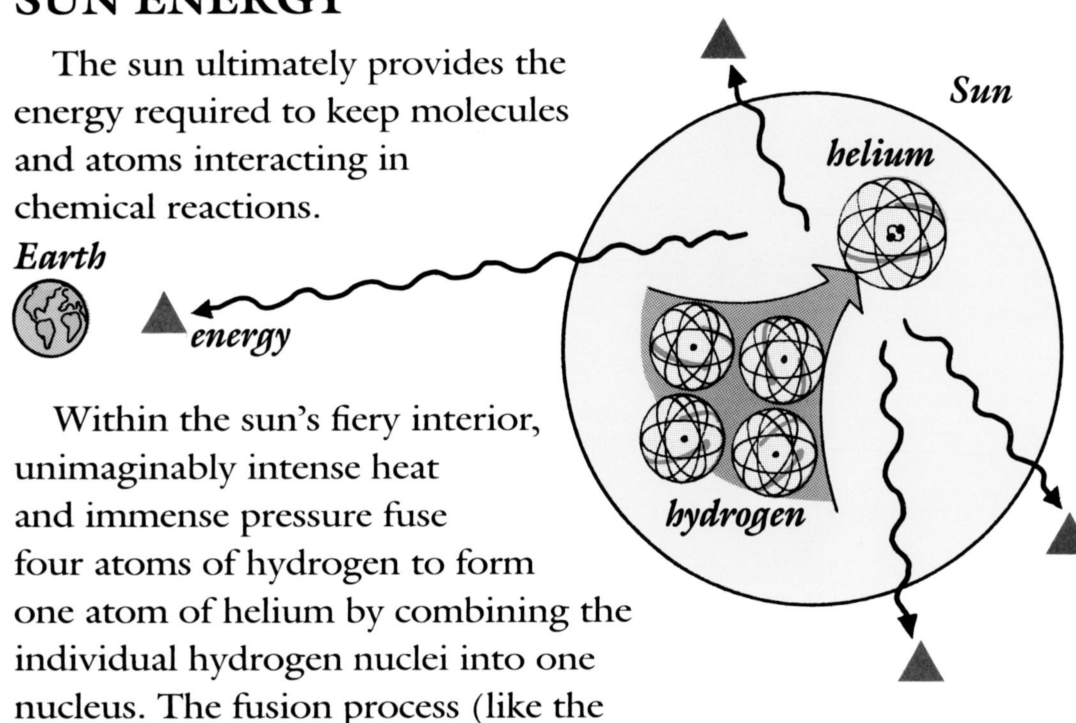

Within the sun's fiery interior, unimaginably intense heat and immense pressure fuse four atoms of hydrogen to form one atom of helium by combining the individual hydrogen nuclei into one nucleus. The fusion process (like the hydrogen bomb) releases tremendous quantities of heat and light energy, which travel to earth across space.

ENERGY TRANSFER

To be useful, energy must be transferred from point A to point B. Molecule A (George) has three options for transferring energy (boiling-hot potato) to molecule B (Sandy). He may hand it to others to pass it on (**conduction**), lob it directly through the air (**radiation**), or run over and hand-deliver it himself (**convection**).

Molecules have the same three choices when transferring energy. **Conduction**: Molecule A passes its energy to other molecules, who transfer it eventually to B, like heat moves up a metal rod from the source.

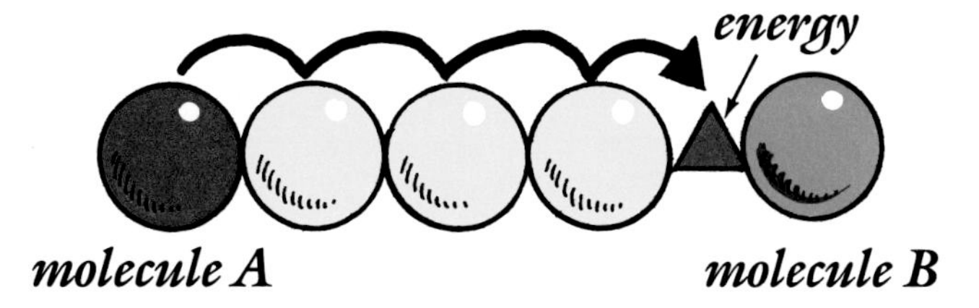

Radiation: Molecule A sends its energy through space to B, like the sun does to earth.

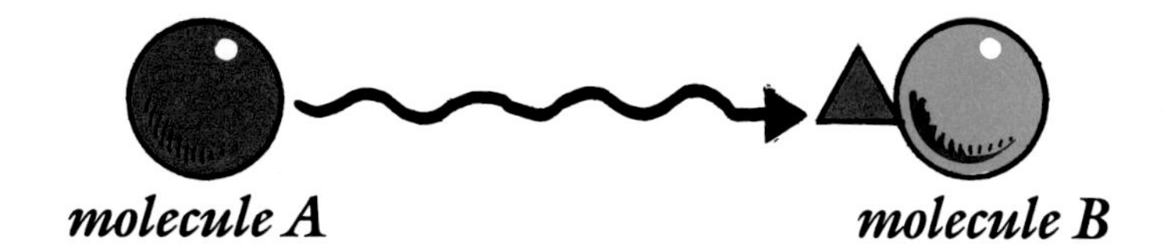

Convection: Energized molecule A carries its energy directly to molecule B for transfer.

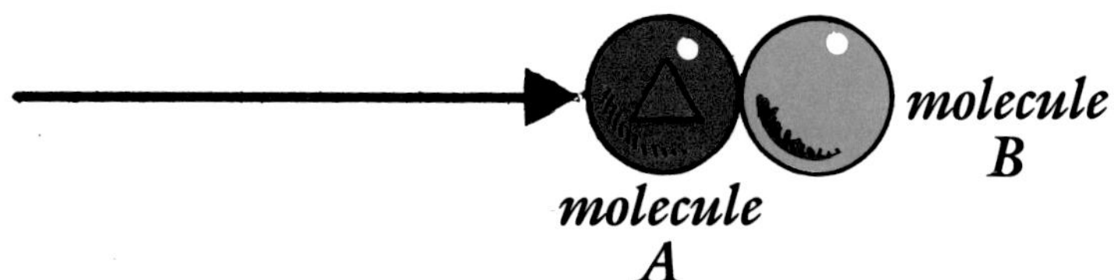

The skier warms her hands by radiation, and the kettle water by **conduction**. Hot gas molecules moving up the stovepipe heat it by **convection**. The pipe then **radiates** its heat.

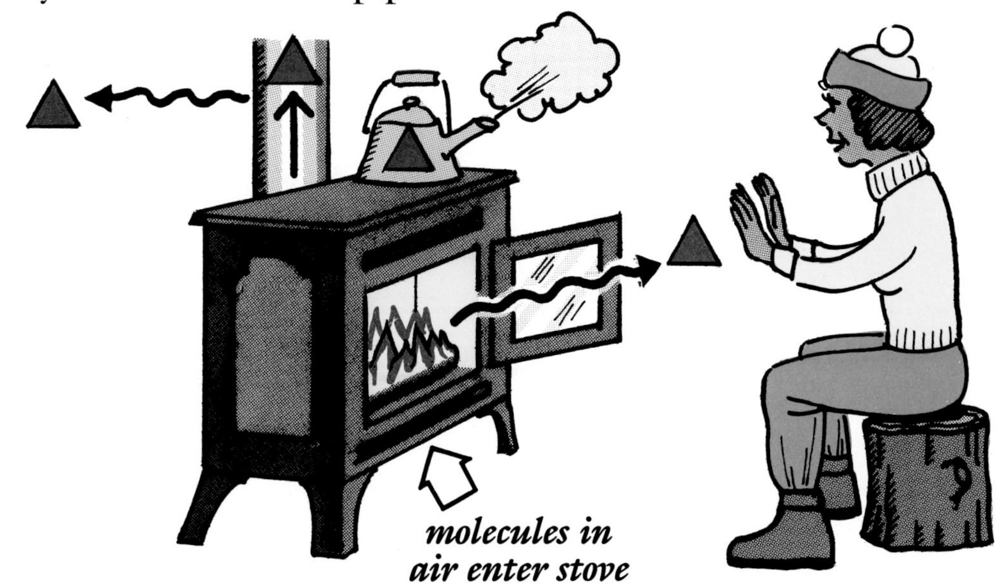

THE PRODUCER REACTION

Green plant life on earth converts the arriving sunlight energy into **food**, a useable and storageable form of chemical energy for present and future requirements. Plants are called **producers** because they manufacture food from atmospheric carbon dioxide. Both plants and animals utilize the food.

Photosynthesis ("made with light") describes the process by which plants manufacture food energy from light energy.

Carbon dioxide molecules and energy from the sun enter the green parts of the plant to engage in a reaction with water molecules from the root system.

carbon dioxide and energy from the sun

water molecules

The Producer Reaction:

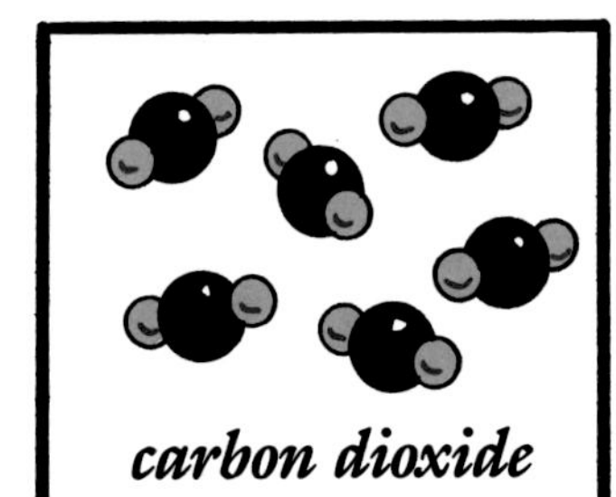

+

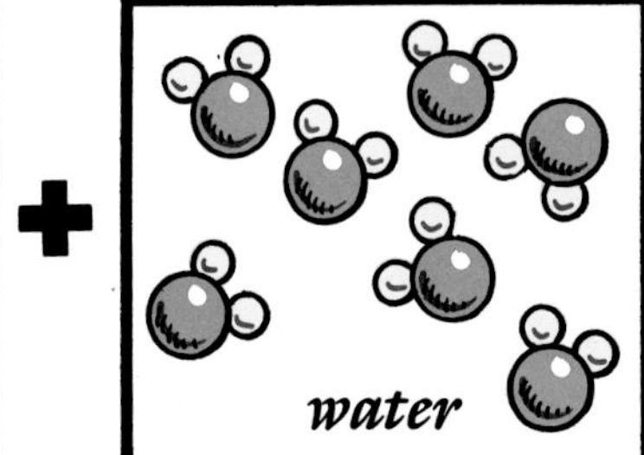

+

A green plant utilizes carbon dioxide gas molecules from the air, water from the ground, and sunlight in a chemical reaction resulting in the assembly of carbon atoms in structures of varying size and shape, like sugar and cellulose. A bit of the sunlight energy is captured and held between the carbon atoms in the various plant products like sugar and fat and cellulose. Oxygen, a by-product, returns to the air.

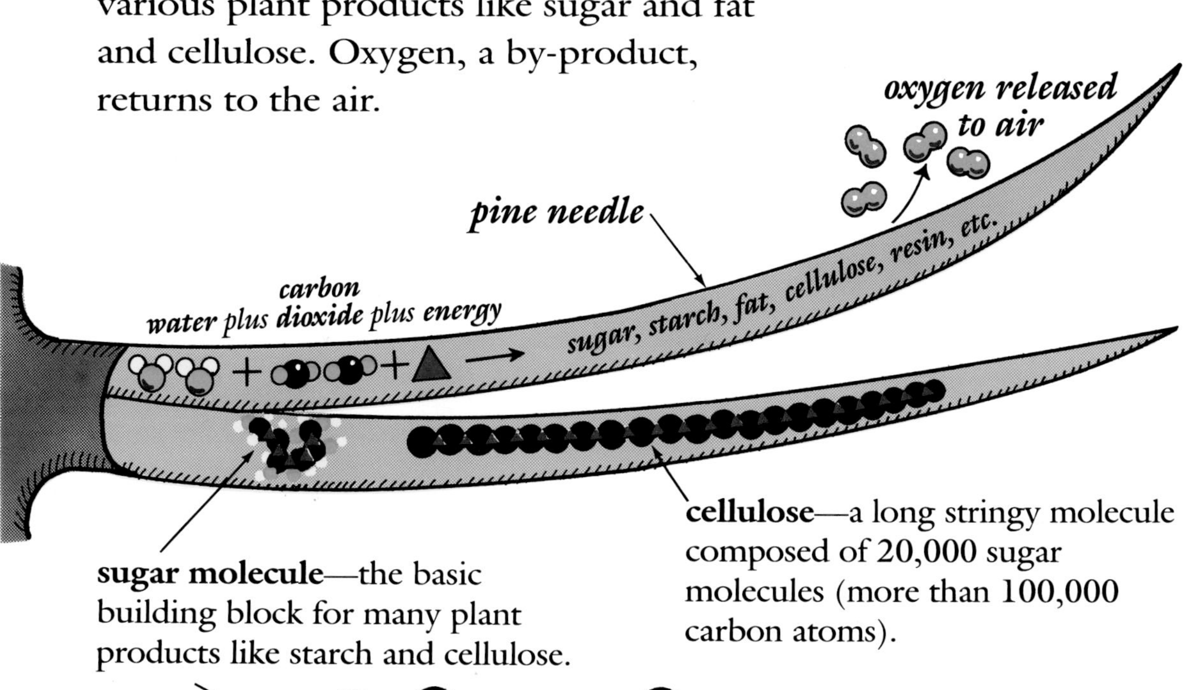

cellulose—a long stringy molecule composed of 20,000 sugar molecules (more than 100,000 carbon atoms).

sugar molecule—the basic building block for many plant products like starch and cellulose.

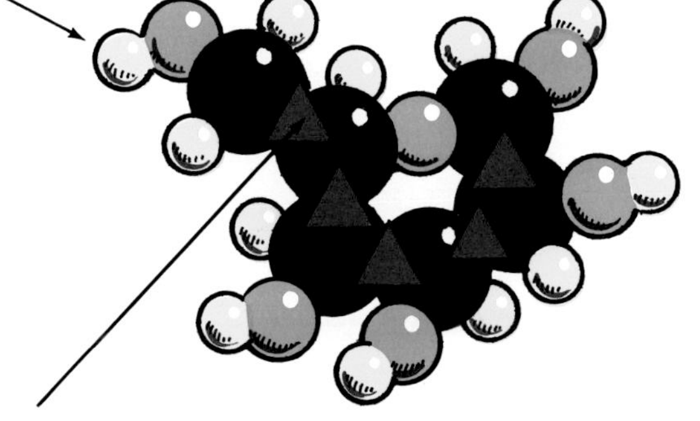

Energy from sun captured between bonded carbon atoms of a molecule of sugar (glucose).

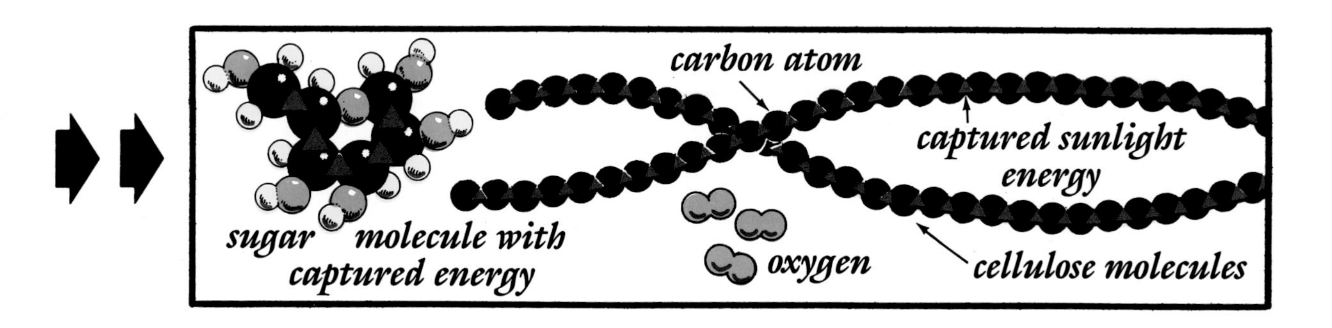

THE CONSUMER REACTION

Animals, directly or indirectly, require the stored food of plants for their energy requirements and are called **consumers**. In order to utilize the stored chemical energy of plants, consumers use chemical reactions that disassemble the carbon-based products that plants originally produced. The reaction releases **chemical energy** stored between the carbon atoms.

The bison's stomach digests the grass and breaks the cellulose into small sugar molecules, which are then combined with inhaled oxygen to release the energy stored in the carbon bonds. The bison uses the released sun energy for all the daily activities of living and exhales the carbon dioxide.

The Consumer Reaction:

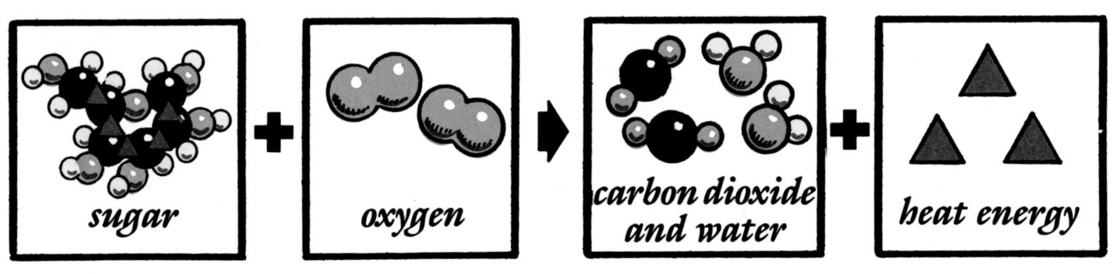

METABOLISM AND COMBUSTION

These campers cooking marshmallows demonstrate the consumer reaction carried out at different rates.

The eager camper on the left will soon carry out the slow version of the consumer reaction and convert the marshmallow sugar to carbon dioxide and water. The energy captured in the sugar molecules will be released very slowly.

The disappointed camper on the right watches the energy stored in the marshmallow sugar release all at once. When all the molecules release their stored energy at once, a fire may result.

Metabolism* = *Slow consumer reaction

Combustion* = *Rapid consumer reaction

The bison uses inhaled oxygen in a chemical reaction that breaks apart the carbon bonds of its stored food molecules of sugar, starch and fat that it obtained by eating **producers**. Released energy between carbon atoms of the stored food heats the bison's shivering body. The bison exhales a moist cloud of breath containing carbon dioxide and water.

The snow machine has similar "food" needs. It breathes in oxygen to mix with carbon chain fuels derived from oil and gas (long-dead producers). A chemical reaction in the cylinder splits the carbon chains apart, releasing the stored energy for the snow machine functions.

The metabolism of food in the bison and combustion of gasoline are essentially the same reaction. Both chemical reactions use oxygen and heat to pry loose the energy stored in the food or fuel. Only the reaction rates vary.

FIRE—A CHEMICAL REACTION

Fire is a rapid and persistent chemical reaction which combines fuel and oxygen to produce heat and light. An external source of heat called the **pilot heat** is usually required to start the reaction.

In an automobile engine, gasoline fuel and oxygen from the air require the heat of an electrical spark to ignite the flames of combustion and start the car.

In natural settings, lightning or volcanos supply the pilot heat to initiate the fire reaction. Like any wild animal, the fire reaction consumes fuel and oxygen for continued existence. If the fire runs out of appropriate fuel, it goes out.

The Fire Reaction:

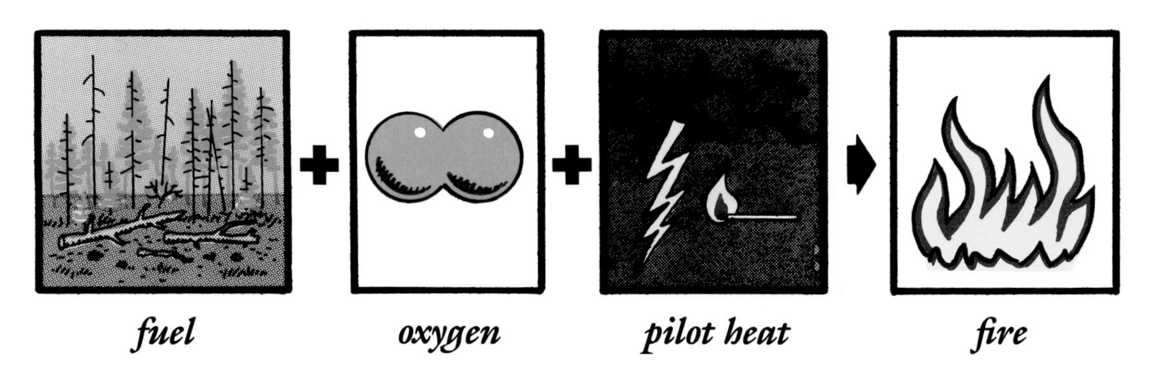

THE FIRE TRIANGLE

The **fire triangle** visually reminds the student of the three requirements to make fire: appropriate fuel, adequate oxygen, and enough heat. The absence of any side of the triangle prevents the formation of fire.

I. FIRE, FLAME, AND COMBUSTION

"Of design, the earth was created to provide its own food from its own waste, and all that it did or suffered was turned back on itself."

—Plato

Fire and **flame** do not share the same meaning. **Fire** is a chemical reaction. **Flame** is the visible manifestation of fire or **flaming combustion**.

Glowing embers display little or no flame and burn by **glowing combustion**. Both types of combustion perform the same chemical reaction. Flaming combustion burns wood-fuel **gases**, while glowing combustion burns wood-fuel **solids**.

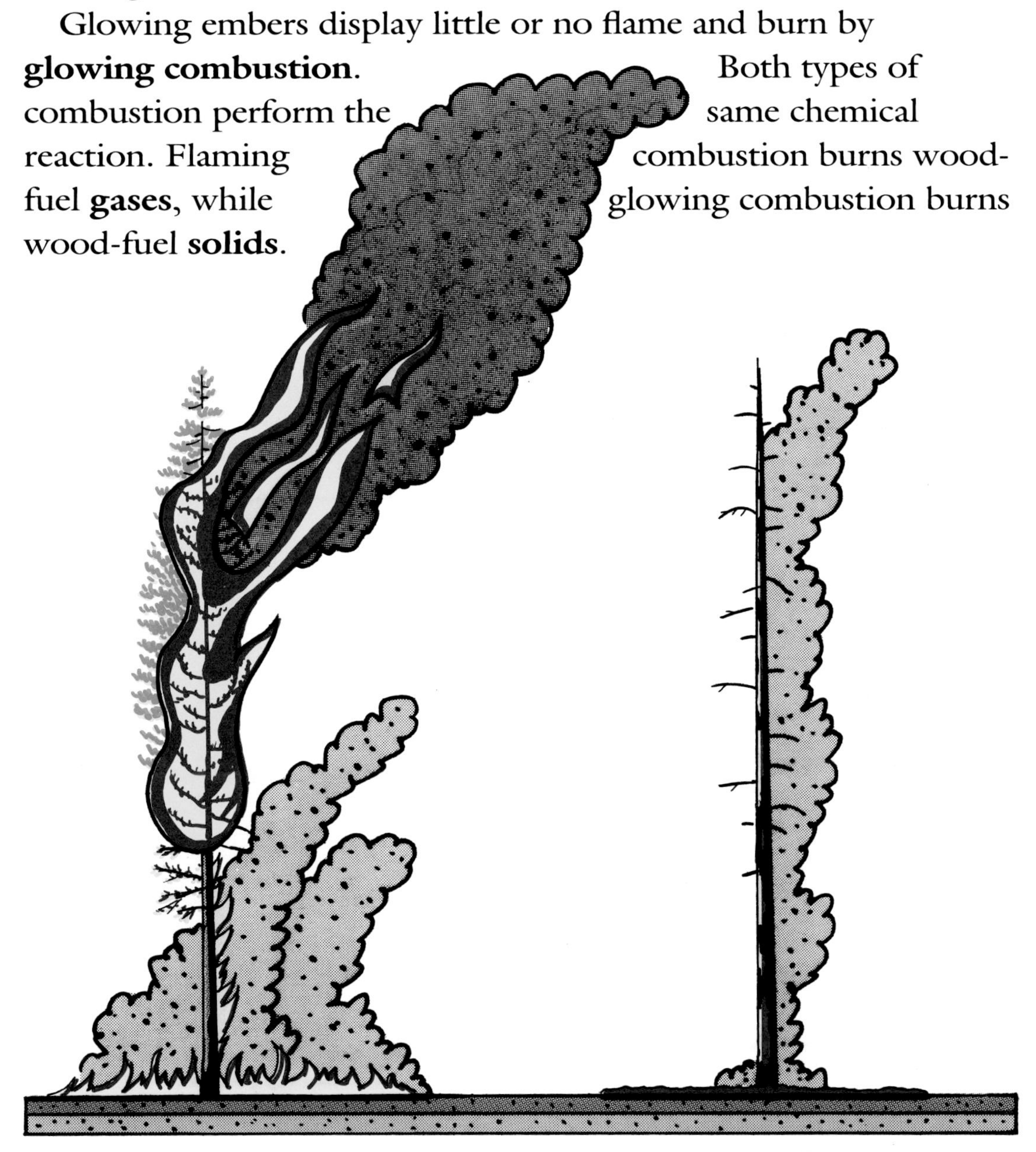

A burning tree demonstrates the transition from **flaming** to **glowing combustion**.

FIRE

When wood burns, it passes through two very distinct phases. A visually athletic **flaming phase** consumes the potential gases stored in the wood. A prolonged and subtle **glowing phase** releases the energy stored between the atoms of solid carbon left after flaming subsides.

During the **flaming phase of combustion**, the woody fuel emits gases that burn with a flame. This **aerial chemical reaction** consumes fuel gas *above* the surface of the wood and appears as flexible flame.

When the wood ceases production of fuel gas, the **aerial chemical reaction** descends to the wood to become the **surface chemical reaction** known as glowing.

Flaming requires gaseous fuel.
Glowing requires solid fuel.

FLAME

Flame may be visualized as a **bag** or **tent** filled with hot fuel gases. Oxygen surrounds the bag's exterior. The walls of the bag provide a common meeting ground for the interaction of fuel and oxygen.

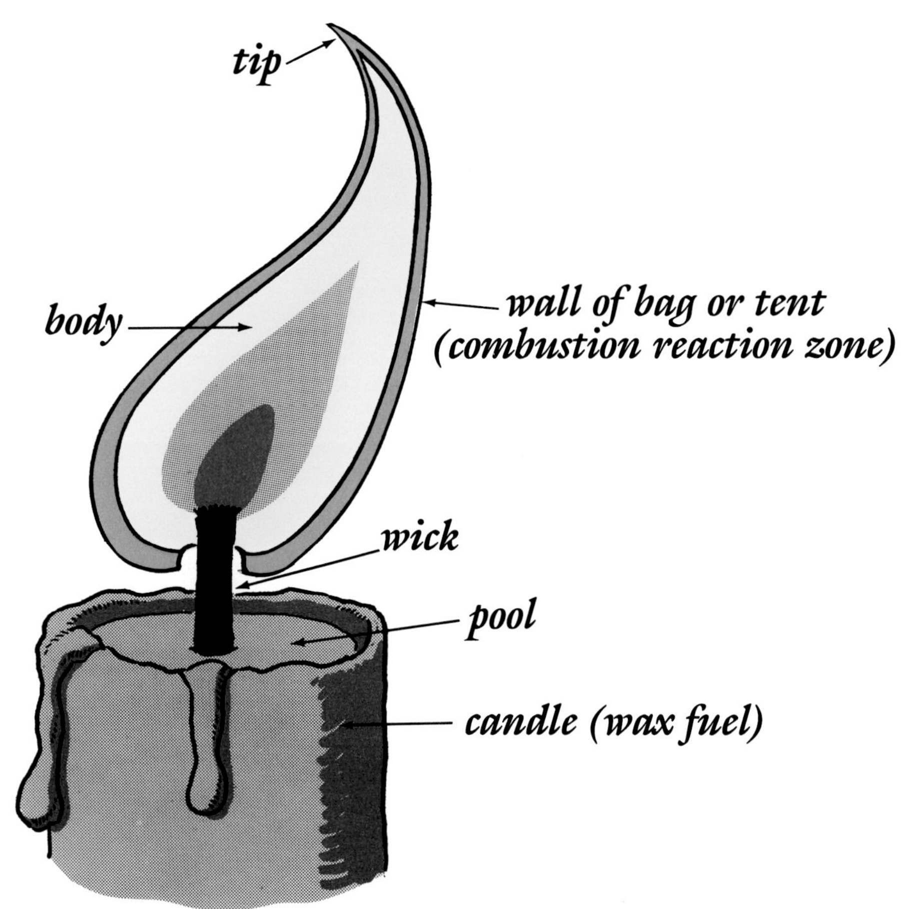

The flexible bag of fuel contains **unburned gases** and **soot**. Within the thin bluish walls of the flame bag, carbon fuel and oxygen combine. This chemical reaction splits the carbon bonds of the fuel to release the stored energy originally acquired from the sun.

Some of that energy heats the soot, causing it to **radiate heat and light** (like a hot light-bulb element).

The bag wall is called the **combustion reaction zone**. If this **reaction zone** fails, the flame smokes and dies. Incredibly complex chemical reactions progress within the flame body enclosed by the reaction zone.

THE STRUCTURE OF FLAME

A burning candle introduces the structure of flame and basic concepts of flaming combustion.

① Fuel (liquid wax) migrates up the wick from the candle pool. It then boils off the wick. Transformed into a fuel gas, it "inflates" the fuel bag and circulates before entering the reaction zone, where it burns.

② Fuel gas molecules enter the fuel bag walls or **combustion reaction zone** and react with oxygen from the surrounding air. Carbon bonds of the stored fuel split and release their energy, which forms the flame walls.

③ By-products of carbon dioxide and water produced in the **reaction zone** float away.

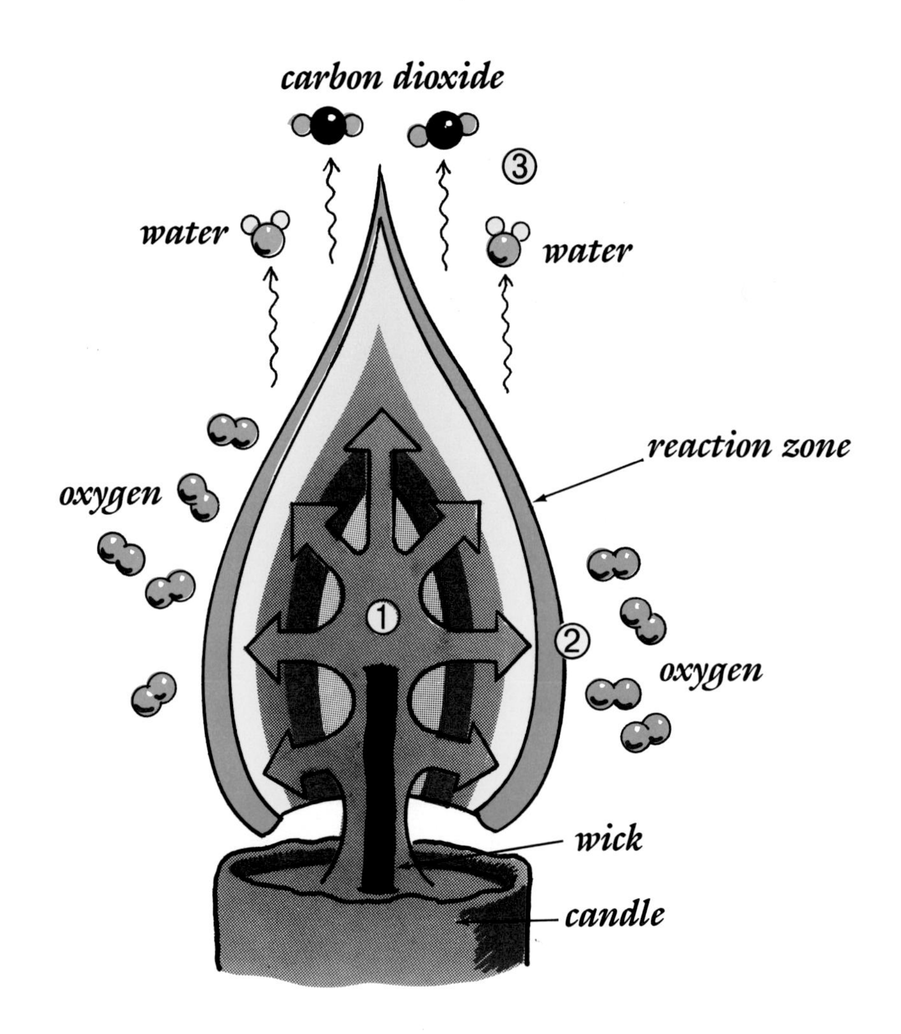

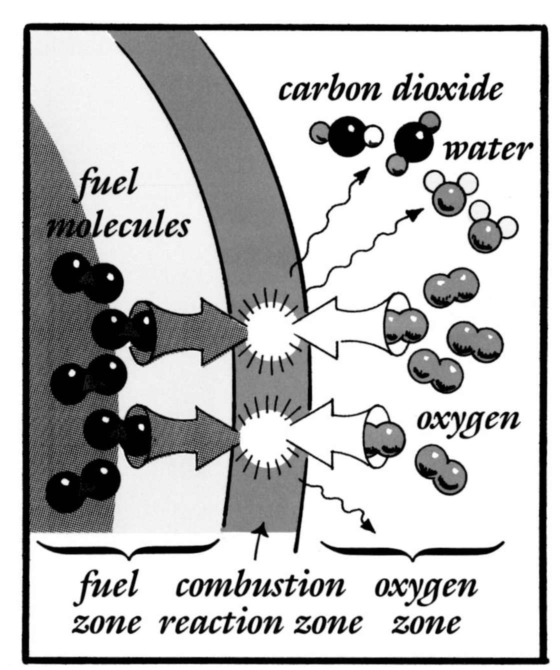

This diagram demonstrates the **combustion reaction zone** and its position, separating fuel and oxygen. Combustion occurs within the "walls" of this zone when carbon-based fuel molecules join with oxygen.

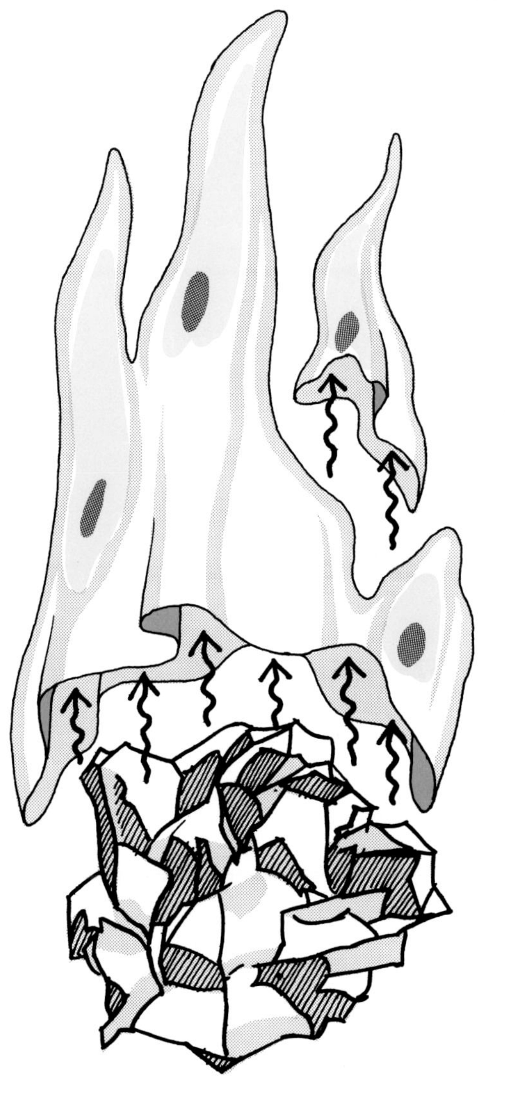

Burning a ball of crumpled paper further demonstrates the flexible bag or tent visual principle. Fuel gases pouring from the many chambers and folds of the crumpled paper inflate the tent into various fantastic shapes. The flame floats on the invisible gas molecules. When the gas in the paper depletes, the tent will collapse onto the paper to continue burning solid fuel.

PYROLYSIS PATH AND SOOT CYCLE

Fuel Behavior Inside of a Flame

① Fuel vaporizes out of the wick and enters the flame's interior as long chains of carbon atoms. The fuel molecules may randomly enter either the **pyrolysis path** or **soot cycle**.

Pyrolysis Path

② Pyro-lysis means "heat-divided." The intense heat radiating from the **reaction zone** does *not* burn, but **subdivides** the fuel chains into smaller and smaller fragments. **Pyrolysis** of fuel makes it easier to burn, like kindling split from a large round of wood.

Pyrolysis reduces the large fuel molecules to fragments of 2 to 4 carbons in length before entering the reaction zone, where they join oxygen to form carbon dioxide. When the carbon bonds break, heat energy from the sun is released, and the by-products of carbon dioxide and water migrate away from the reaction zone.

③ Flame accompanies this reaction of fuel gas combustion.

Soot Cycle

④ Some fuel molecules avoid entering the reaction zone by collecting themselves together into snowball-like particles called **soot**. These conglomerations of fuel fragments and other chemical substances from the circulating brew in the flame's interior disperse themselves throughout the flame body.

Soot particles, like light-bulb filaments, radiate light of varying color and intensity, depending on the amount of wattage or energy they absorb. Soot absorbs energy from the reaction zone and radiates a yellowish to orange light, depending on proximity to the reaction zone. Soot near the wick absorbs little heat, radiates little or no light, and appears black.

Radiating soot is responsible for the yellowish colors of flame. Burning fuel in the reaction zone yields a delicate blue.

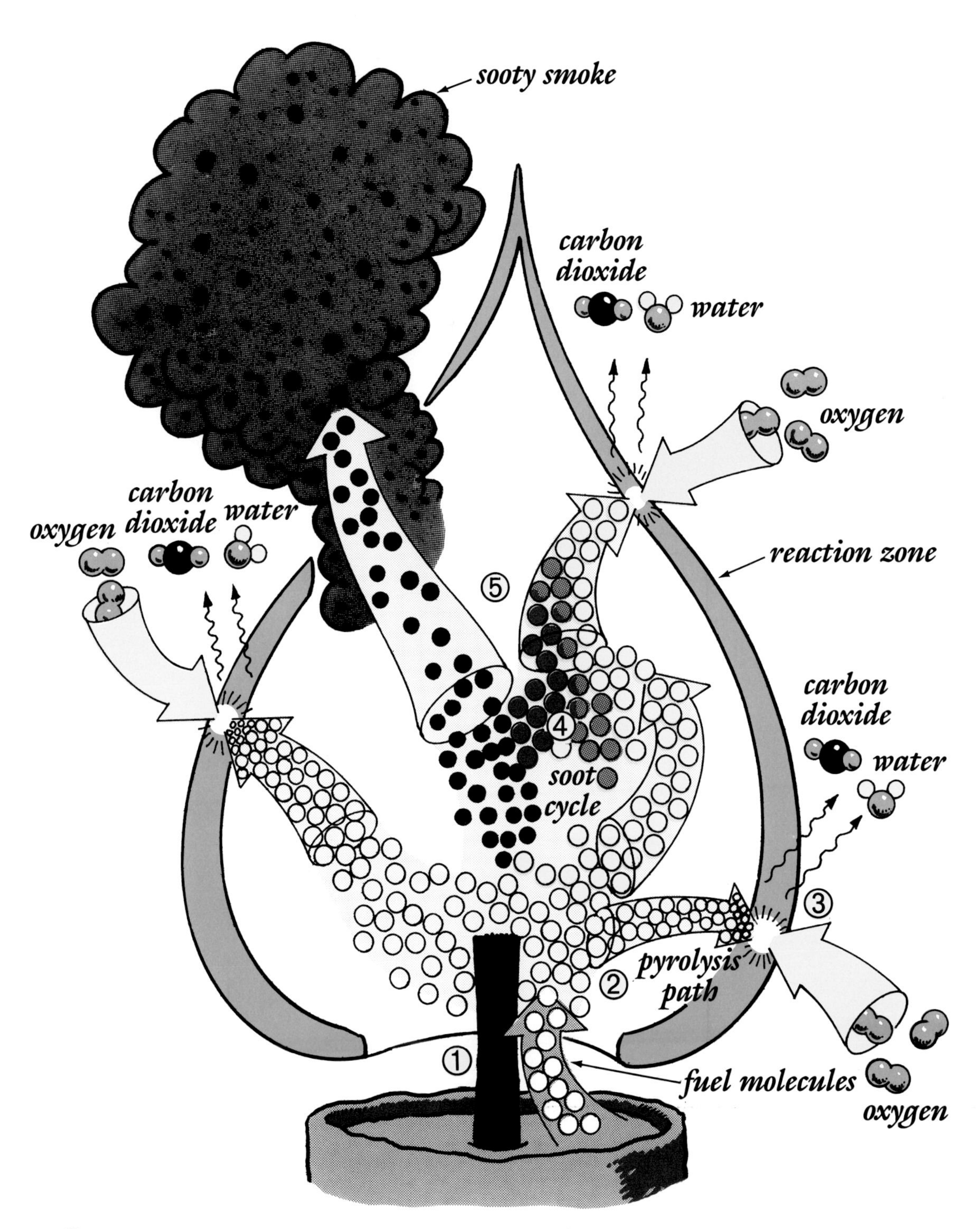

⑤ The soot particles may then enter the reaction zone and be combusted to carbon dioxide and water. They may also exit the flame body through an inefficient or absent section of reaction zone and immediately cool to blackness as soot smoke. Flames generate black sooty smoke when their combustion reaction zones fail to burn all the fuel and soot completely.

FLAME IGNITION

The **fire reaction** requires pilot heat to start making fire. In a forest, pilot heat originates from some external source like lightning, chain saw sparks, neglected camp fires, cigarettes, and matches. Ignition depends on external sources of energy.

Pilot-dependent Fire Reaction:

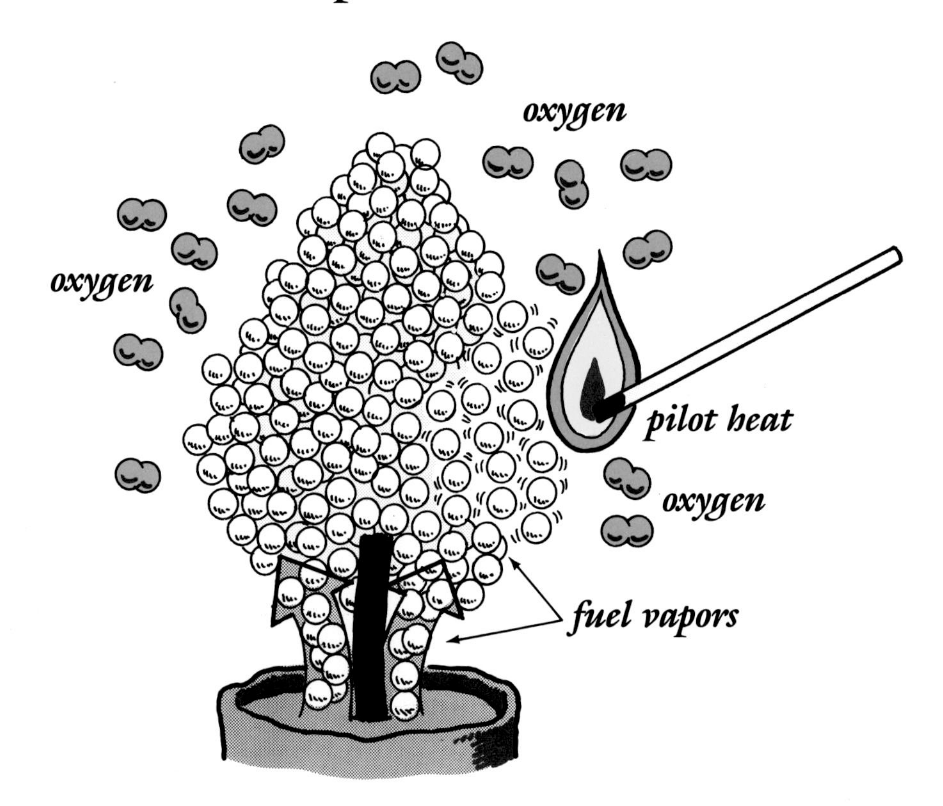

Ignition of flame combustion requires gaseous fuel. In this candle example, the match flame melts and vaporizes the wax to form a **cloud** of flammable fuel gas around the wick.

Molecules of oxygen and fuel react most rapidly near the pilot flame, whose heat increases the reaction speed.

Eventually, enough fuel molecules split and release their combined stored energy to initiate a continuous reaction. Molecules start exploding apart like a book of burning matches. The bluish combustion reaction zone suddenly bursts to life. The zone spreads rapidly in all directions from the pilot heat until it encloses the cloud of fuel vapors with its tent.

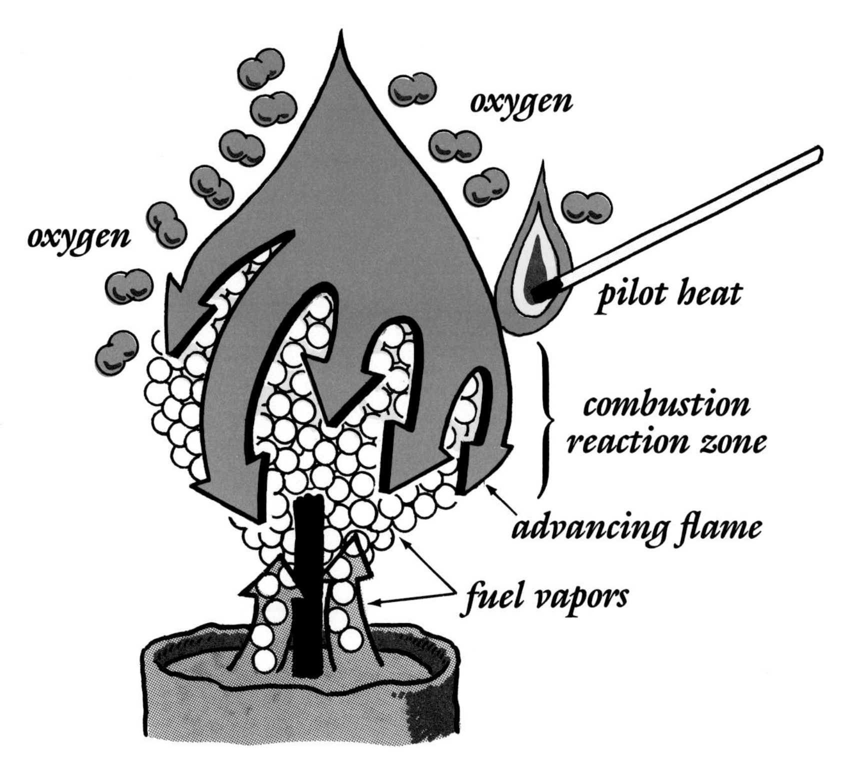

The reaction zone soon produces enough heat to sustain its own reactions without the match's heat. The reaction zone now supplies the heat for the fire reaction independent of a match.

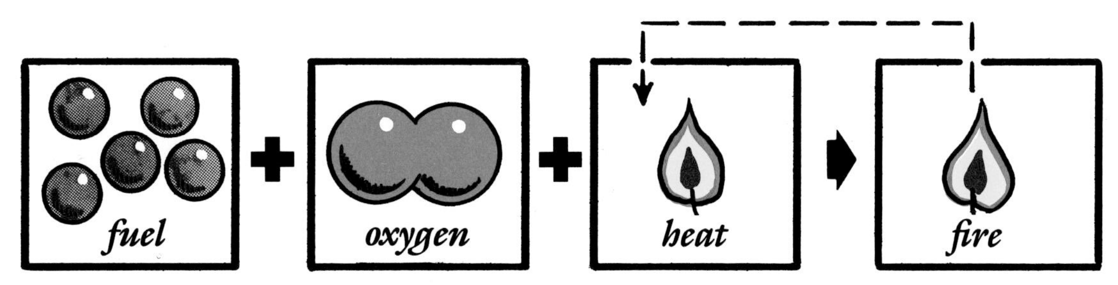

Independent Fire Reaction:

Flaming combustion requires gaseous fuel, which travels to the combustion reaction zone. **Solid fuel burns in the opposite manner**. The reaction zone must travel to the fuel.

FUEL PYROLYSIS

Woody forest fuels burn much differently than wax candles. Pyrolysis of wax molecules yields a finely divided gaseous fuel for flaming. Pyrolysis of woody fuel yields a gaseous fuel *and* a solid fuel, which burns in a distinctively different manner.

Pyrolysis is the most important but least appreciated concept in forest fuels combustion. Everyone is acquainted with **pyrolysis**, which describes the physical alteration of fuel by heat.

Defective toasters deliver charred bread in a cloud of tar smoke. The fuel in the bread entered **pre-ignition** and produced a fuel cloud. The bread never actually burned, but **scorched**, due to fuel pyrolysis.

A hot iron may alter the cellulose structure of cotton fabrics, resulting in a permanently charred surface and a puff of tar smoke. The fabric scorched but never burned.

The cigarette smoker inhales the tar smoke from scorched tobacco cellulose or tobacco cellulose pyrolysis.

CELLULOSE PYROLYSIS

When the extremely long molecules of cellulose absorb scorching heat, they physically break into smaller molecules, which rebond to each other to form a surface of **char** and a smoke of **tar**. This heat-induced physical process is called **pyrolysis**.

Tar gas and droplets contribute greatly to the fuel gas cloud and to pyrolysis-smoke formation.

Char is the layer of black carbon atoms that remains after the tar gas evolves and leaves the heated cellulose surface.

Char is the solid component of forest fuels.

Char and **tar** contain many high energy bonds that contribute heat energy when released by combustion.

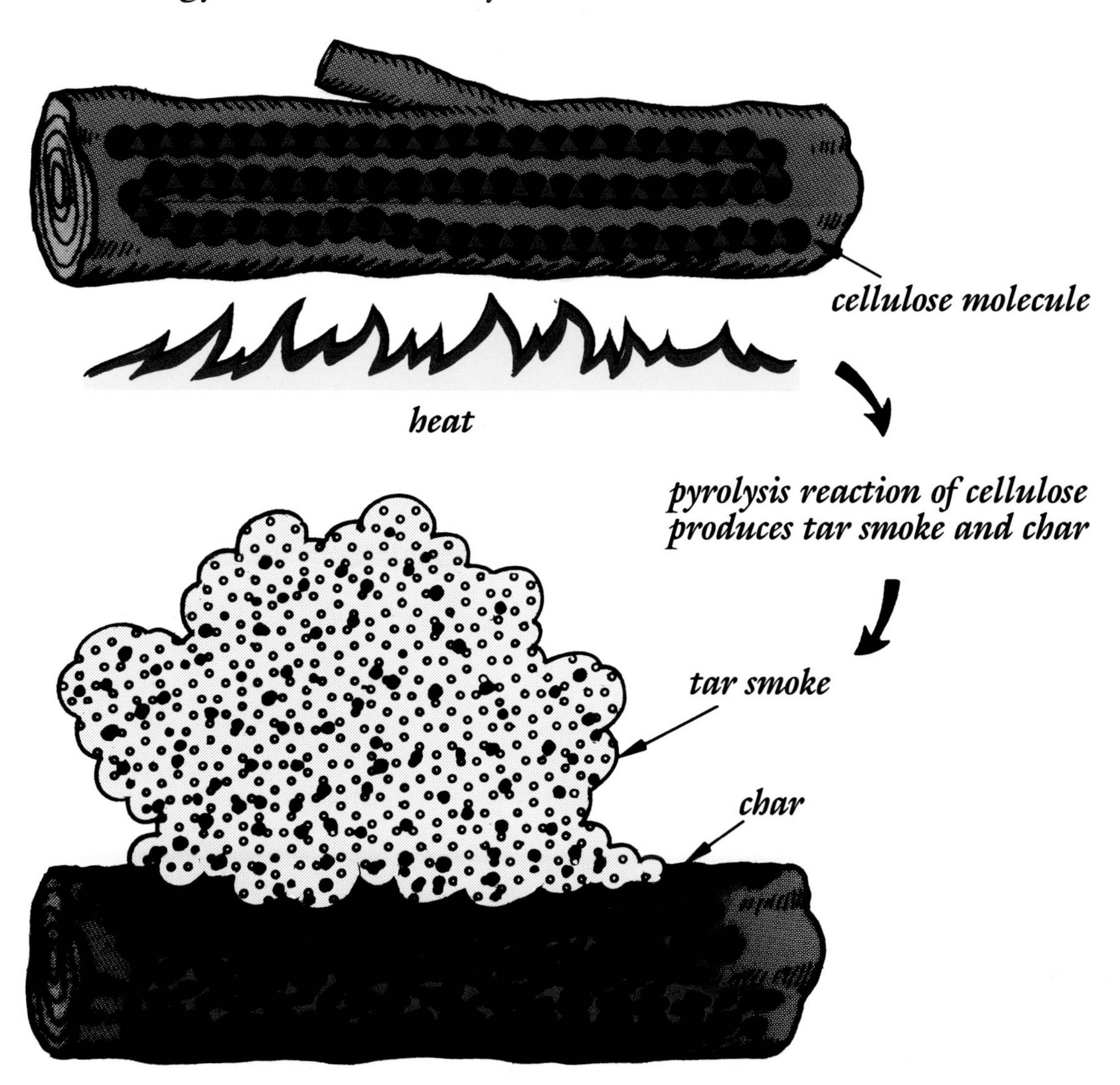

PHASES OF COMBUSTION

All forest fuels pass through the three basic phases of combustion: **pre-ignition**, **ignition**, and **combustion**. Many forest fuels burn initially by flaming combustion of fuel gases, followed by glowing combustion of fuel solids.

Solid fuel combustion is best appreciated by observing the manner in which a piece of typical forest fuel burns, and observing the transition through the various stages.

PRE-IGNITION PHASE

Pre-ignition dehydration occurs as the pilot heat dries the fuel by boiling off the water in the area of applied heat. Pilot heat must be adequate in duration or intensity to dry the fuel, or the pre-ignition phase stops. Rain-soaked wood prolongs this phase, while drought-dried woody fuel dramatically shortens this phase.

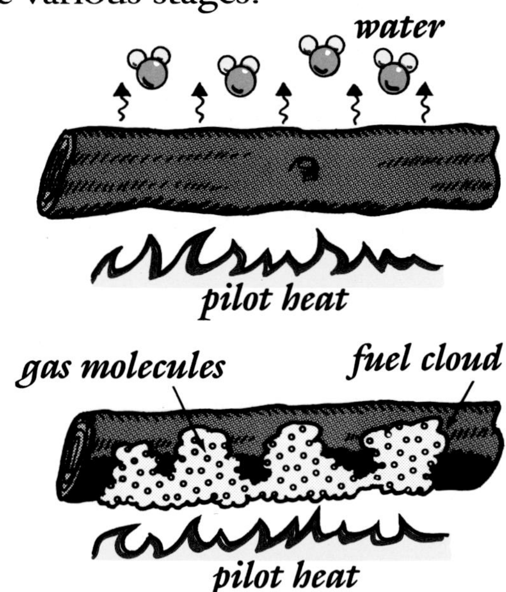

Pre-ignition fuel cloud formation follows continued pilot heat application. The fuel absorbs more energy and the temperature elevation outgasses (boils off) substances in the wood like fat, oil, turpene, alcohol, and resin, which collectively form a cloud of fuel molecules.

Persistent heat application **scorches** the fuel, producing a blackened surface of **char** that emits a gray smoke of **tar**. Tars in gas and condensed droplet form contribute greatly to the fuel gas cloud.

Scorching describes the heat-induced physical alterations of the fuel surface and the gases it produces. **Scorching is cellulose pyrolysis**.

fuel gas cloud
tar droplets
pilot heat

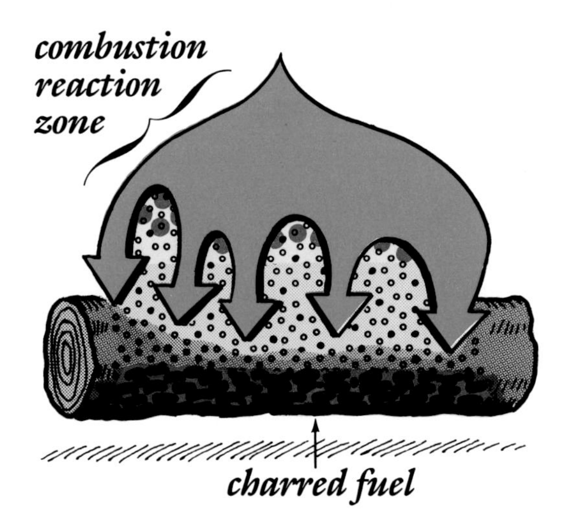

IGNITION PHASE

The **combustion reaction zone** materializes and rapidly encloses the fuel cloud. The fire now burns independently with the pilot heat removed.

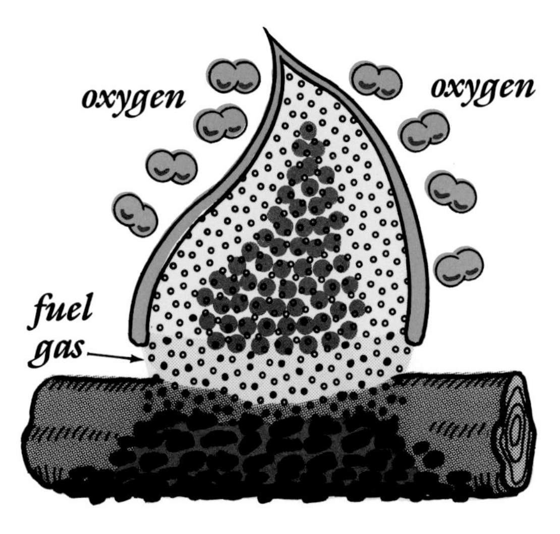

COMBUSTION PHASE

The radiant heat of **flaming combustion** dislodges the fuel gas molecules which enter the fuel cloud and eventually burn. Radiant heat also preheats and pre-ignites unburned wood.

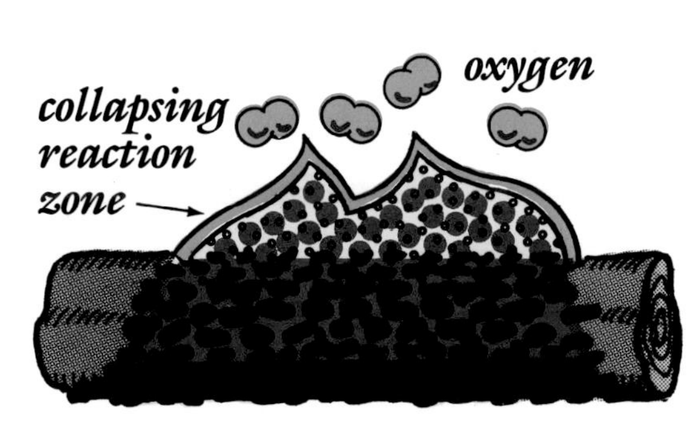

The wood eventually runs out of fuel gas after all the cellulose pyrolyzes to char and tar, and the fuel cloud grows smaller without the continuous flow of wood fuel gas. The tent of flame dwindles and collapses like an improperly pitched tent.

The collapsing reaction zone spreads out on the charred wood and initiates **glowing combustion**.

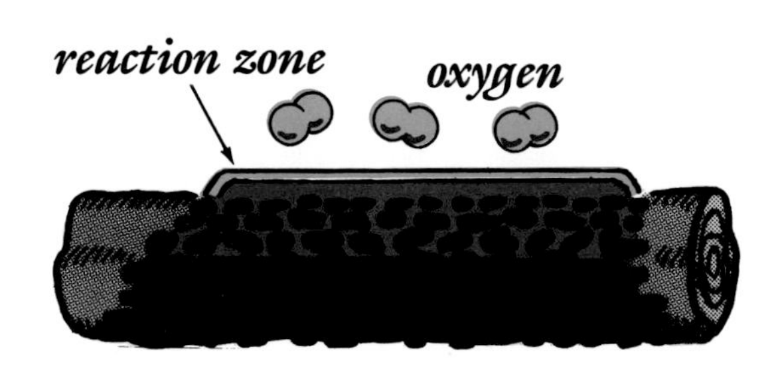

The reaction zone must travel to the fuel in order for combustion to continue.

GLOWING COMBUSTION

When the combustion reaction zone collapses onto the charred surface of the wood fuel, it settles like a heavy mist into every nook and cranny of the charred network of cellulose carbons. The reaction zone continues to provide a common meeting ground for the immobile carbon fuel and atmospheric oxygen.

The reaction zone heats the underlying carbons to reddish-orange glowing embers. This is the beginning of **glowing combustion**, which will gnaw away at the solid carbon fuel molecules locked into the surface of charred cellulose and lignin (the molecule that stiffens the cellulose in woody plants). The reaction zone heats the molecules. The vibrating molecules disrupt their connecting bonds and release their stored energy.

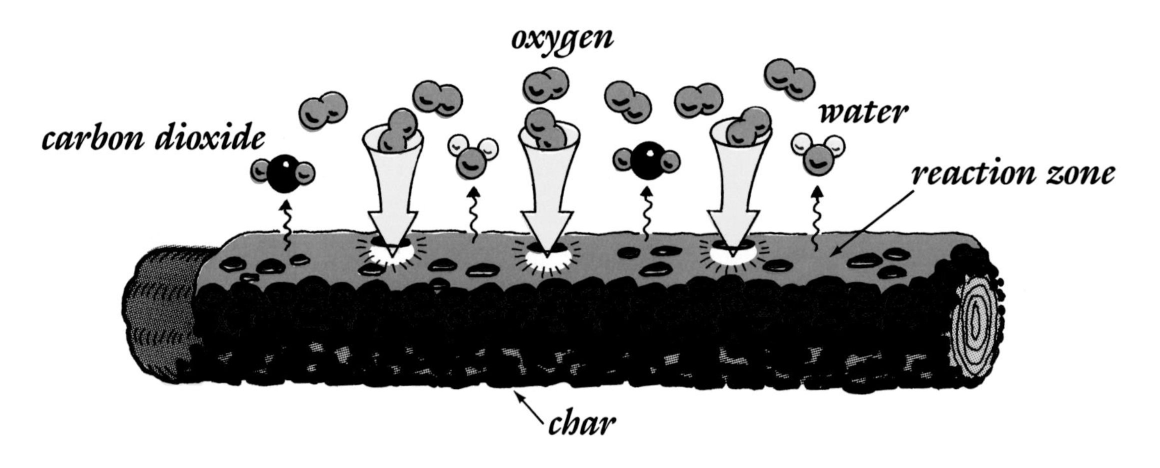

In actuality, the charred surface inhibits the movement of oxygen into the fuel, which results in the production of carbon monoxide (CO), a deadly poisonous gas. When this gas emerges from the fuel, it joins another oxygen, burns with a blue flame, and becomes carbon dioxide (CO_2).

In fuel-rich, oxygen-poor environments (e.g. underground roots, dead log interiors, peat, under ashes) the deadly CO cannot burn. This flameless or **smoldering combustion** claims human lives yearly.

Glowing combustion eats away into the carbon lattice-work of the pyrolyzed cellulose and lignin molecules, converting them to carbon monoxide, carbon dioxide, and water.

The heat of glowing combustion causes the release of pre-ignition phase smoke in adjacent unburned wood. The heat of glowing fails to ignite the fuel gases and the scorching wood adjacent to the glowing carbon fuel produces large volumes of tarry smoke of pyrolysis.

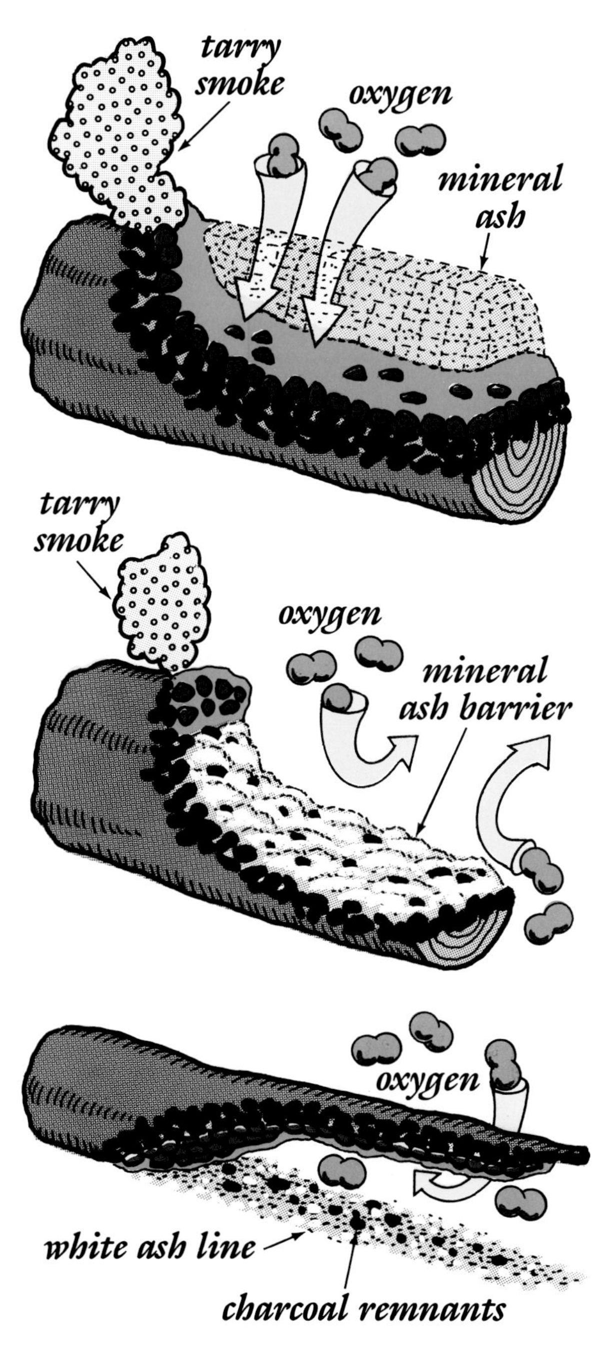

After the reaction zone burns away all the carbons in the fuel, only a delicate network of whitish mineral ash remains. The ingredients of ash resemble those of sand and do not burn.

Mortar has little strength after the bricks are removed. Likewise, after carbon atoms leave the wood, ash has no strength and collapses to form an **ash barrier** that may block oxygen from getting to the reaction zone. The reaction zone goes out, as does the glowing combustion under too thick a layer of ash.

In usual downed forest fuel, glowing combustion burns from the bottom to top because the mineral ash falls away from the reaction zone, thereby giving oxygen easier access to the carbon. Stirring and repositioning wood in a camp fire knocks away the smothering ash barrier.

PHASES OF COMBUSTION—SUMMARY

Pre-ignition phase—dehydration and fuel cloud formation.

Pilot heat dries the fuel and then forms a flammable fuel cloud composed of flammable wood gases and tar droplets derived from the pyrolysis of cellulose. The fuel surface chars.

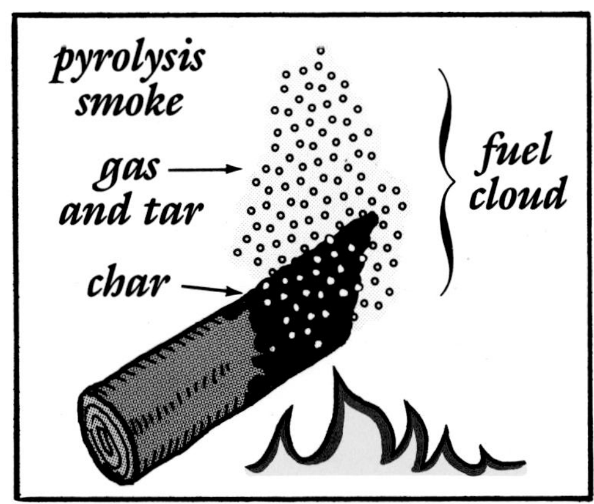

Ignition phase.

The combustion reaction zone forms about the fuel cloud and the pilot heat is no longer required.

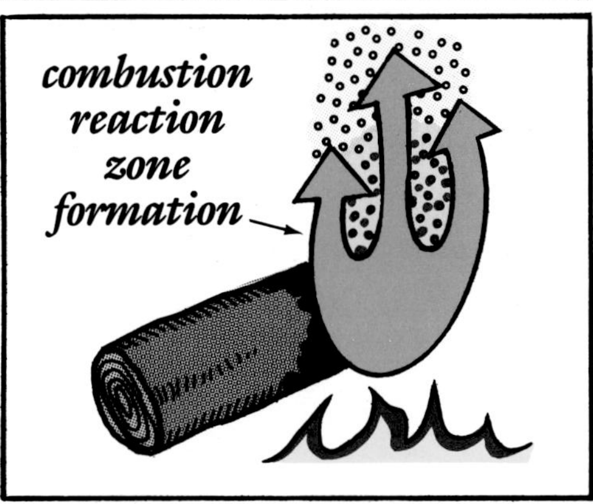

Combustion phase—flaming and glowing.

Flaming combustion consumes the gaseous fuel produced by wood. The flames preheat the adjacent unburned wood, which emits a fuel cloud of pre-ignition. The fire spreads when this cloud ignites. Flames dominate this phase of combustion.

Glowing combustion attains phase dominance after the wood ceases fuel gas production. Glowing combustion releases the immobile carbon atoms locked in the fuel.

Glowing continues until it releases all the fuel carbons or it is extinguished by smothering ash.

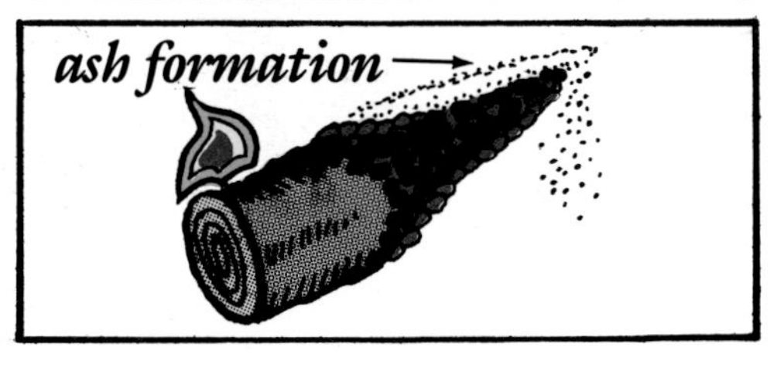

III. FIRE IN FORESTS

"Unless one comes to an understanding concerning the nature of change, one will have many difficulties." —Plato

Forest fuel, **terrain**, and **weather** determine the type and behavior of individual forest fires. The principles presented in earlier chapters readily transfer to predicting how a twig, tree, or forest burns.

In this section, illustrations define the various layers of the forest and forest floor. The location and character of the forest fuel help determine if fire burns the forest or the forest floor. Numerous determinants of fire behavior are introduced, but the applications of the phases of combustion to forest fuels is emphasized. The section ends with smoke.

Backing Fire:

FOREST LAYERS

A wildland forest consists of a plant community dominated by trees growing on a forest floor.

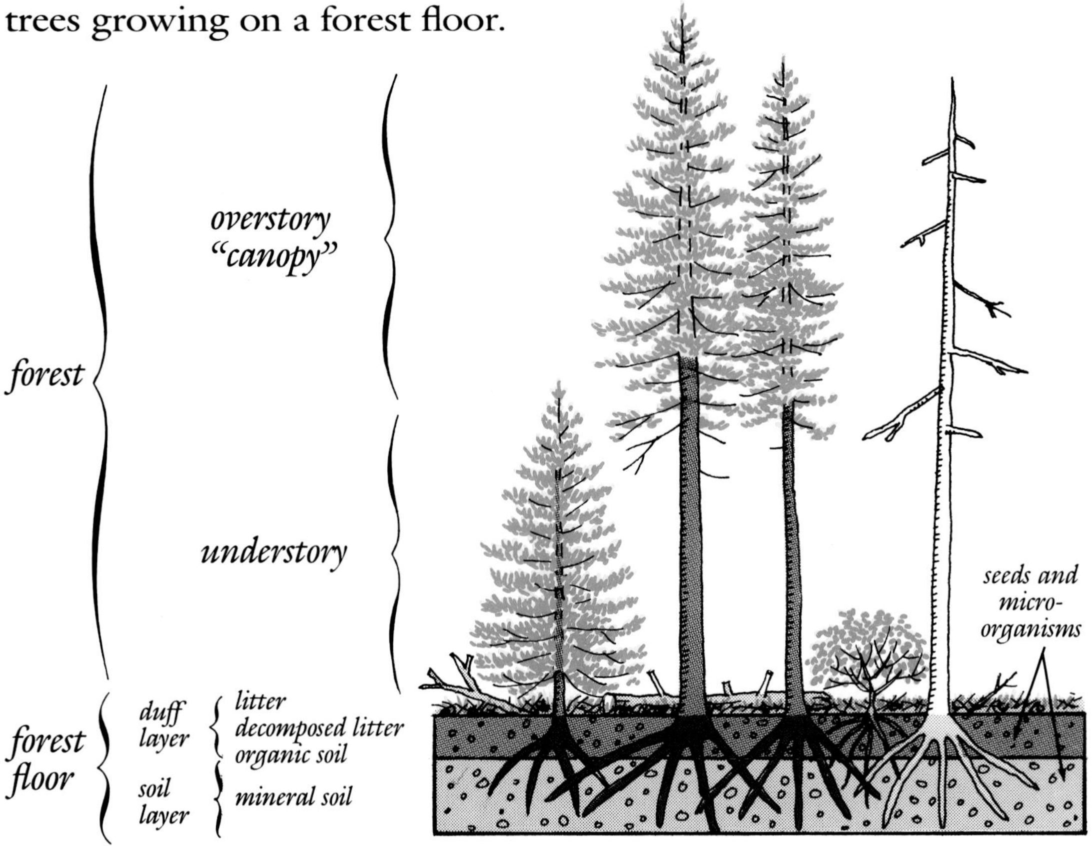

The **duff layer** resides upon the **soil layer**. Duff evolves from each year's dead and decomposed plants and animals. Mineral soil evolves from decomposition of rocks like granite or rhyolite, sandstone or basalt.

The duff and upper mineral soil **both** contain roots and seeds, bacteria and fungi. Burrowing worms and moles mix layers.

FOREST FUELS

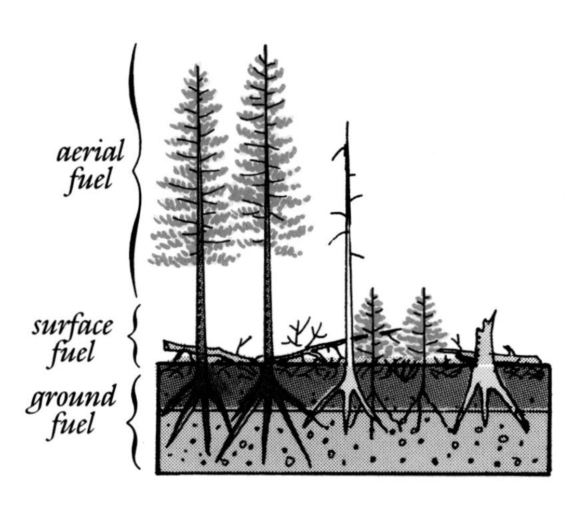

Any substance that will ignite and combust (burn) is a fuel.

Combustible material higher than one meter (39 inches) above ground level is **aerial fuel**. **Surface fuel** includes all combustibles less than one meter above ground level and one year's litter. **Ground fuel** includes all combustible substances below the surface litter of the duff.

FUEL CHARACTERISTICS

① Dead or living—more than half the weight of living fuels is water used by the plant for biological processes.
② Dry or wet—fuel must be dry to burn.
③ Fine or heavy—fine fuels measure less than 1/4 across and kindle and carry fire to heavy fuels.

FIRE TYPES

Most wildland fires are complex blends of **ground fire**, **surface fire**, and **crown fire**.

ground fire in ground fuel *surface fire in surface fuel* *crown fire in aerial fuel*

FUEL POSITION AND FIRE BEHAVIOR

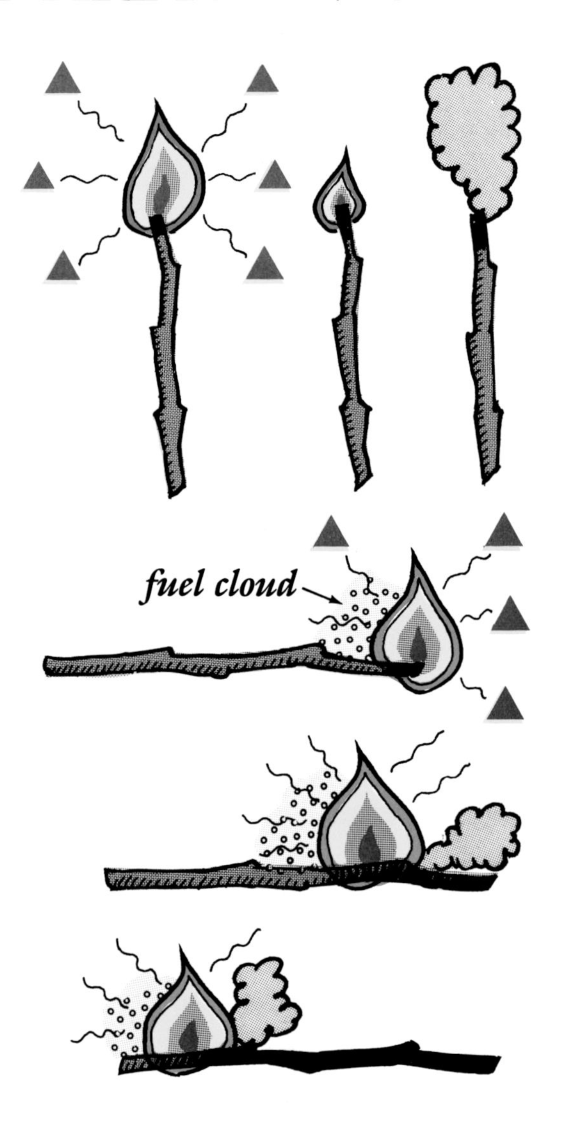

A burning twig clearly illustrates how fuel position determines the spread of fire.

In the vertical position, a twig quickly fizzles out. Very little flame energy radiates to the unburned portion of fuel. The flame cannot preheat the adjacent unburned twig, which fails to complete the first phase of combustion, **pre-ignition**.

In the horizontal position, the flame is closer to some of the unburned twig, which it more easily radiates and preheats. The preheated wood generates a flammable fuel cloud and completes the **pre-ignition phase**. The existing flame serves as a generous pilot heat for the **ignition phase** in the new fuel cloud. The **combustion phase** follows quickly, and the three phases spread the fire slowly but surely along the twig.

In the inclined position, the radiating flame body is much closer to the unburned wood, which it intensely preheats.

The distance between flame and unburned fuel greatly affects the rate of fire spread.

These principles apply readily in a forest setting. In a forest fire, unburned fuels absorb heat from burning fuels. The transfer of heat from burning fuel to unburned fuel occurs by radiation and convection *(see pages 8–9)*.

The distance between flame and fuel determines the amount of energy absorbed by radiation. When the flame and fuel are in close proximity to each other, fuel absorbs *much* more energy.

In convective transfer, the flame must reside beneath the unburned fuel so hot molecules can rise from the region of flame to the wood and preheat it.

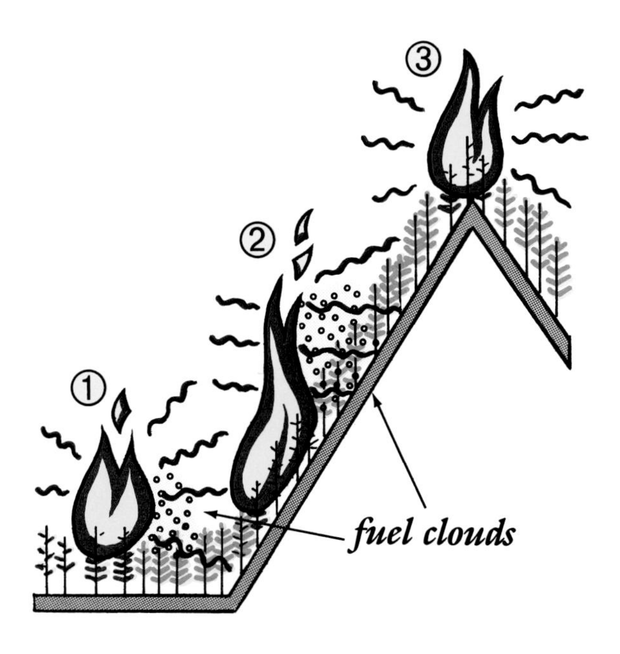

Flame ① preheats (by radiation) only a few trees at any time. The fire spreads slowly. Flame ② preheats many more trees because of its position on the hillside. The fire spreads much more rapidly and "runs" up slope. Flame ③ radiates no trees and rapidly uses up existing fuel and dies out. Many natural fires die on sharp ridges and mountain tops.

Wind pushes the flames closer to the unburned trees, which allows the flames to rapidly radiate and preheat them. Uphill fire movement brings the flames closer still to the unburned trees. Wind encourages downhill fire by bending flames close to fuel that might have escaped preheating.

Flame ① preheats very little fuel and may not ignite the scattered ground fuel. Flame ② preheats a lot of fuel and will progress rapidly through the three phases required for fire. This flame preheats unburned wood by radiation *and* convection.

HOW A TWIG WITH NEEDLES BURNS

Pre-ignition phase—dehydration and fuel cloud formation.

A pilot heat source dries the fuel by boiling off stored water. Then a fuel cloud develops, containing molecules derived from hot sap, resin, turpene, oils, and tar droplets of pyrolyzed cellulose. The twig and needles scorch, producing tar and char.

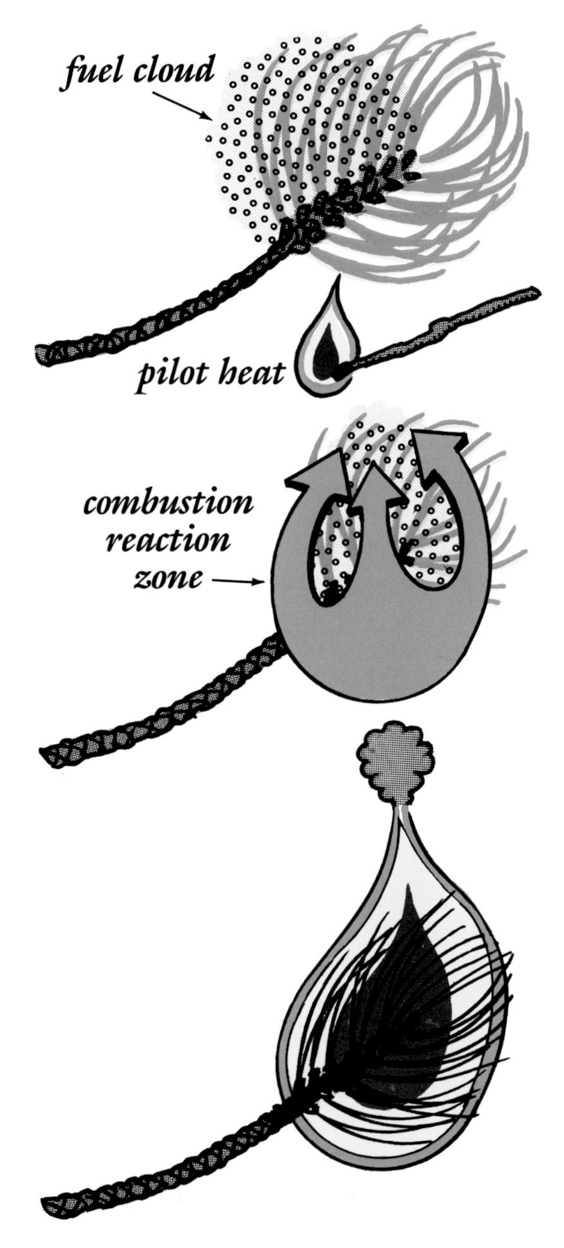

Ignition phase.

The pilot source ignites the flammable fuel cloud. A combustion reaction zone spreads immediately from the ignition point to envelop the entire fuel cloud and establish continuous combustion.

Combustion phase—flaming.

Flaming combustion rapidly depletes the tiny supply of fuel gases released by the twig and needles.

Combustion phase--glowing.

Glowing combustion takes over after flaming consumes the twig's gas production, and leaves a solid fuel of charred needles and twigs.

Glowing combustion slowly releases all the remaining carbon and its energy, leaving only a white mineral ash.

The fragile skeleton of mineral ash awaits disturbance.

The delicate ash debris drifts to the ground under gentle conditions, or blows away with the wind. Strong updrafting wind may carry the ash hundreds of miles from the fire.

HOW A TREE BURNS

Torching describes the astonishing process of nearly instantaneous combustion of a tree in a few seconds. Very hot and dry climatic conditions promote torching. Torching occurs on windless days and in isolated trees or small groves of trees. When the intensely hot combustion reaction zone of flame actually touches a fine fuel like pine needles, the phases of combustion shorten drastically.

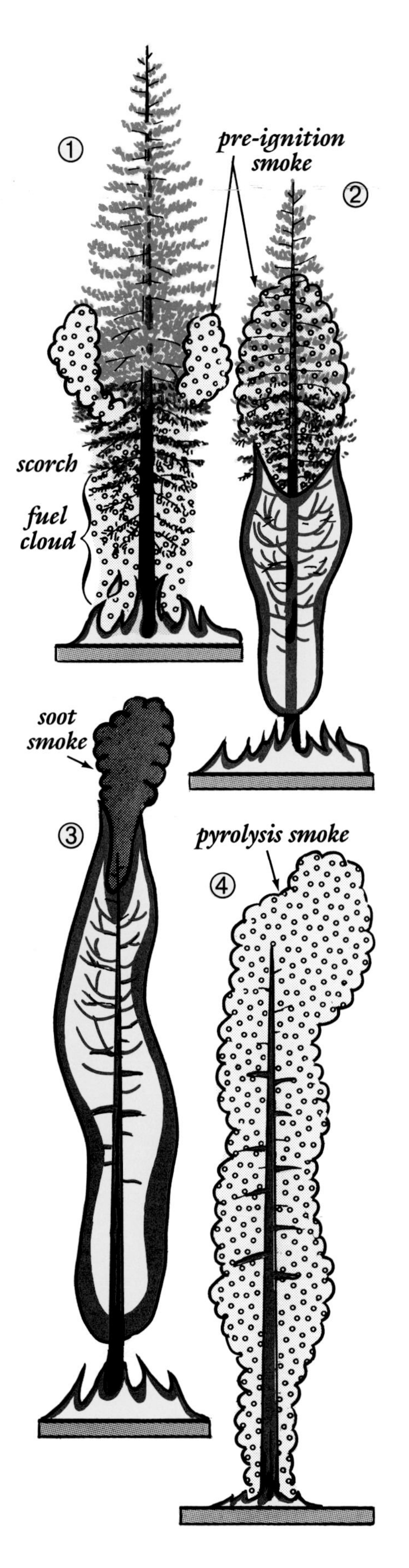

① Burning surface fuel preheats the lower tree. **Pre-ignition phase** terminates with a hot fuel cloud, scorched bark and needles, and tarry smoke.

② **Ignition phase** explosively consumes the fuel cloud and envelops the lower tree in flames. The upper tree instantly preheats and enters the pre-ignition phase.

③ **Flaming combustion** envelops the entire tree and its fuel gas cloud.

④ **Glowing combustion** persists in the charred surfaces left by flaming combustion, but rapidly dies out in living trees because of high water content. The flaming fuel cloud scorches the entire trunk black, but much of the bark remains recognizable.

WIND EFFECT AND TWO-TONES

When surface winds blow strongly, the fuel cloud, caused by surface fire preheating, concentrates on the downwind side of the tree.

The concentrated fuel cloud ignites.

The flaming fuel cloud climbs the tree trunk and scorches the wood bark.

The two-tone trunks reveal which way the wind blew when they burned. Two-tones contain mixtures of dead brown needles, scorched black needles, and even living green needles spared by the blowing wind.

HOW A FOREST BURNS

The majority of forest fires start from the pilot heat supplied by lightning or by man's matches or sparks. The fire first burns in the surface and ground fuels. If fuel and atmospheric conditions of low moisture and humidity permit, the fire may climb into the tree tops or canopy of the forest, utilizing various "ladders."

Fire makes this move in four ways:

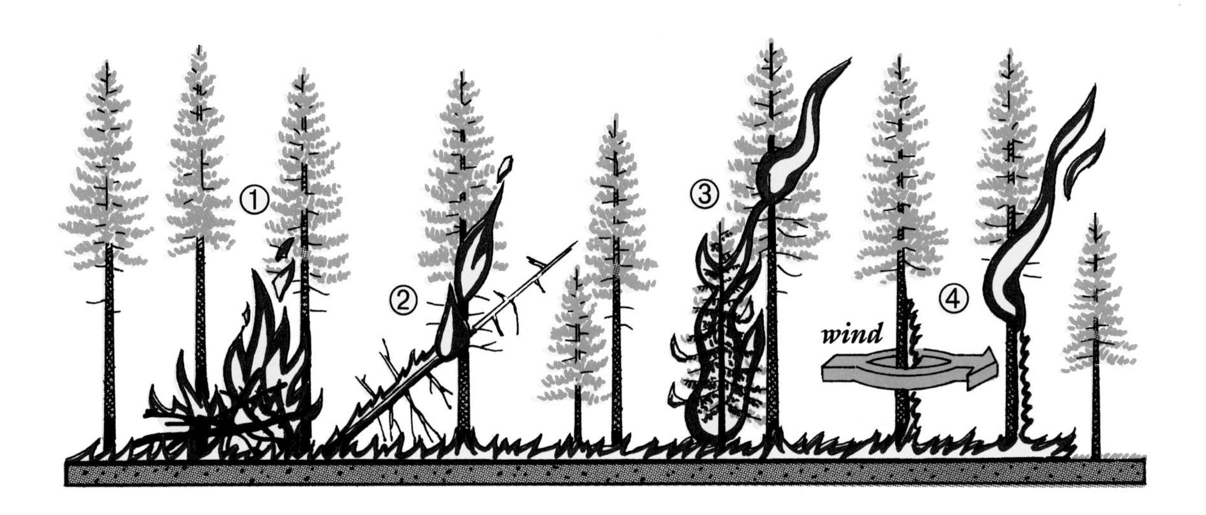

① Plentiful dry surface fuel burns hot enough to preheat and ignite the canopy of aerial fuel.

② Fire works up a partially-fallen tree to the canopy.

③ Surface fire ignites an understory tree whose branches drape near to the ground, yet reach up to the canopy.

④ If the living tree water content is very low, as in a drought, and there is a wind, the fuel cloud concentrates downwind of the trunk, ignites, and flames "climb" up into the crown.

When the fire gains access to the tree crowns, it is called a crown fire. Surface and crown fires cooperatively preheat unburned trees in advance of the moving flame front. This **coupled** relationship is called a **dependent crown fire**.

When the wind-dependent crown fire races ahead of the surface fire and no longer requires it to assist in fuel preheating, it is called an **independent crown fire**.

This type of **crown fire** may "make a run" of several miles in a single day.

Spot fire describes a fire started by a firebrand (aerial burning fuel) carried ahead of the main fire by wind. The firebrand ignites a surface fire, which may climb into the crown using various fuel ladders.

The fire spreads from the **point of ignition** in three directions called **fronts**. The **heading fire front** rapidly spreads in the direction of the wind. The **backing fire front** slowly creeps against or into the wind. The **flanking fire front** spreads at right angles to the wind.

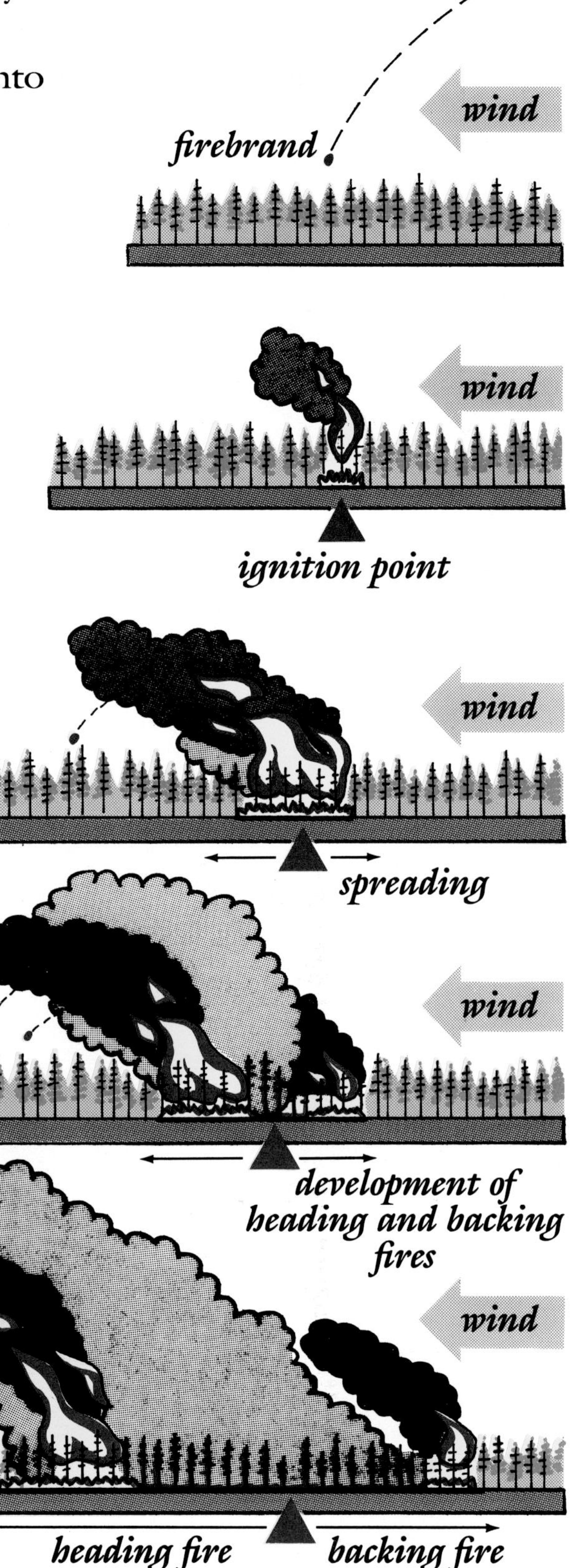

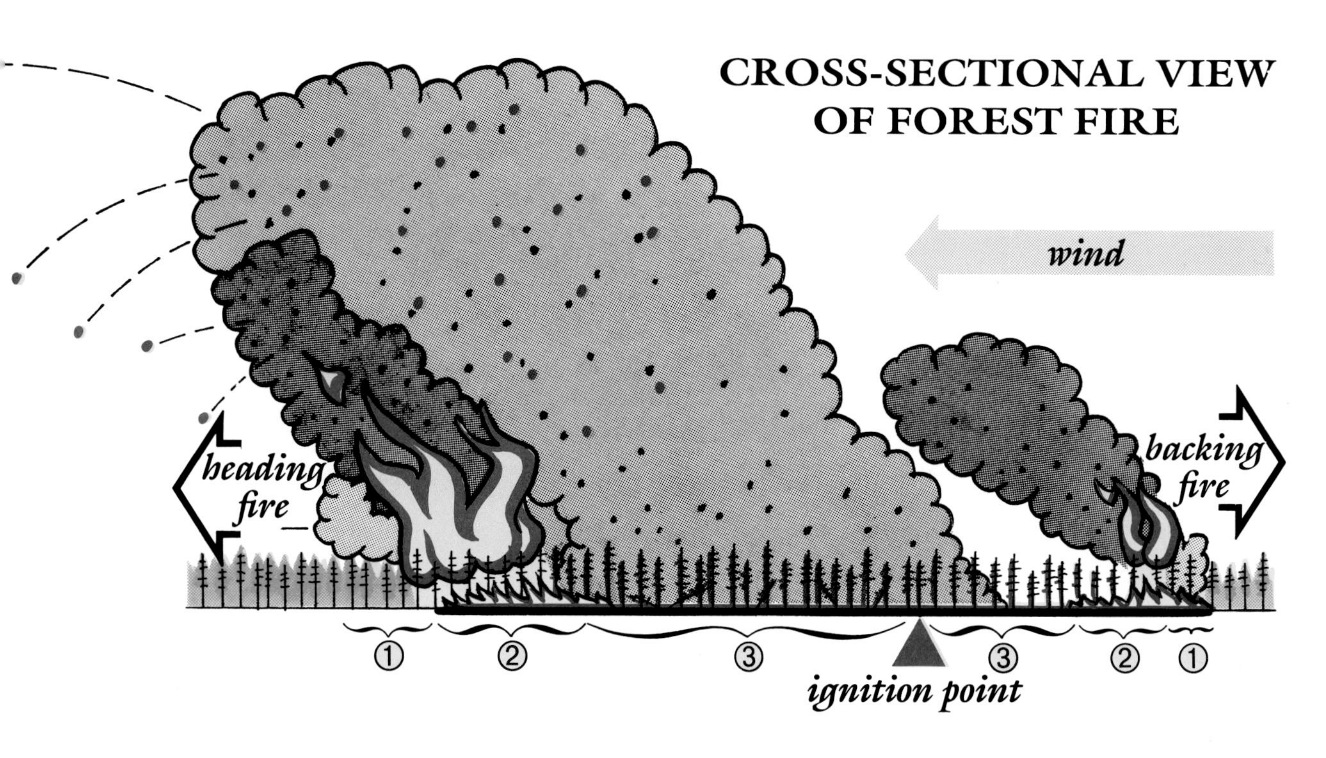

Typical forest fire, showing **heading** and **backing fires**, their direction of movement, and their phases of combustion.
① pre-ignition ② ignition and flaming combustion
③ glowing combustion

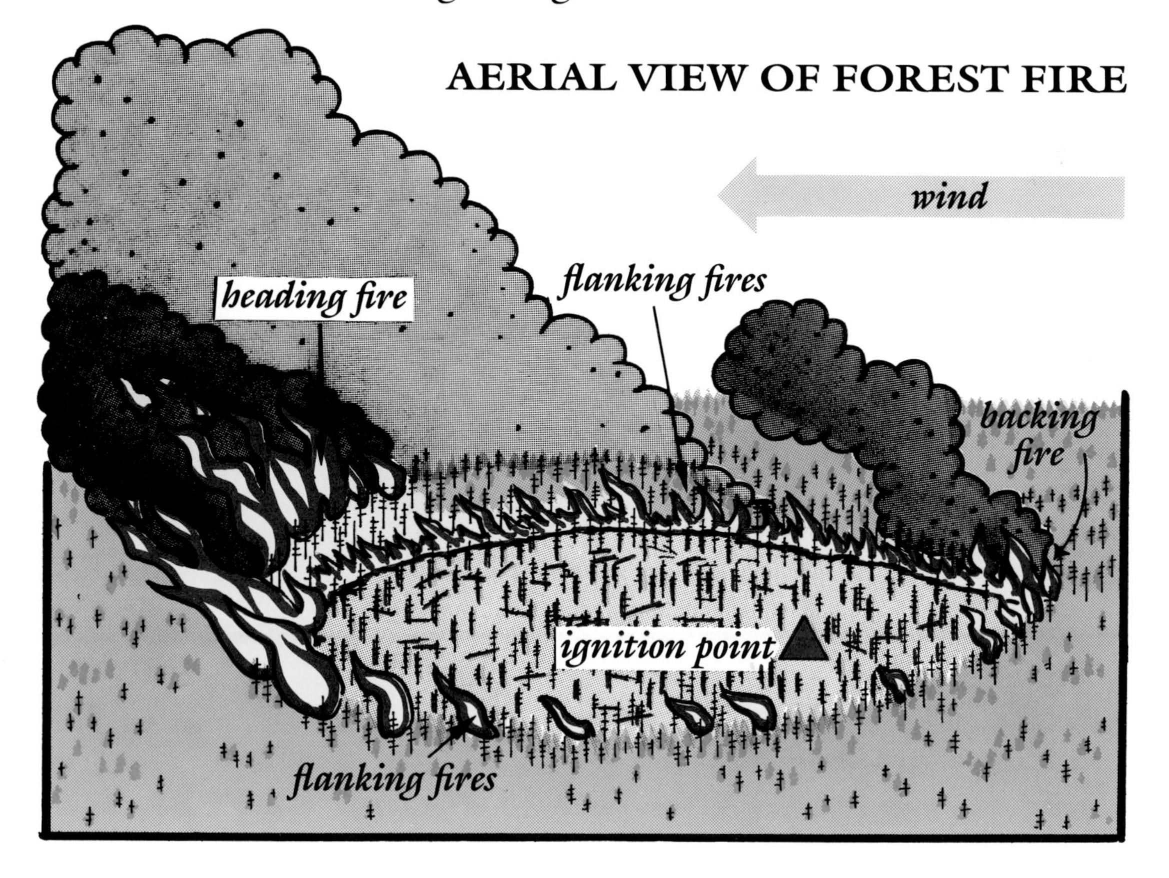

SNAGS—WIDOW MAKERS OF THE FOREST

A **snag** is a dead tree lacking most needles and branches. It burns in several ways. A smoldering ground fire burns out its roots, leaving it teetering, but with no visual clues as to its stability.

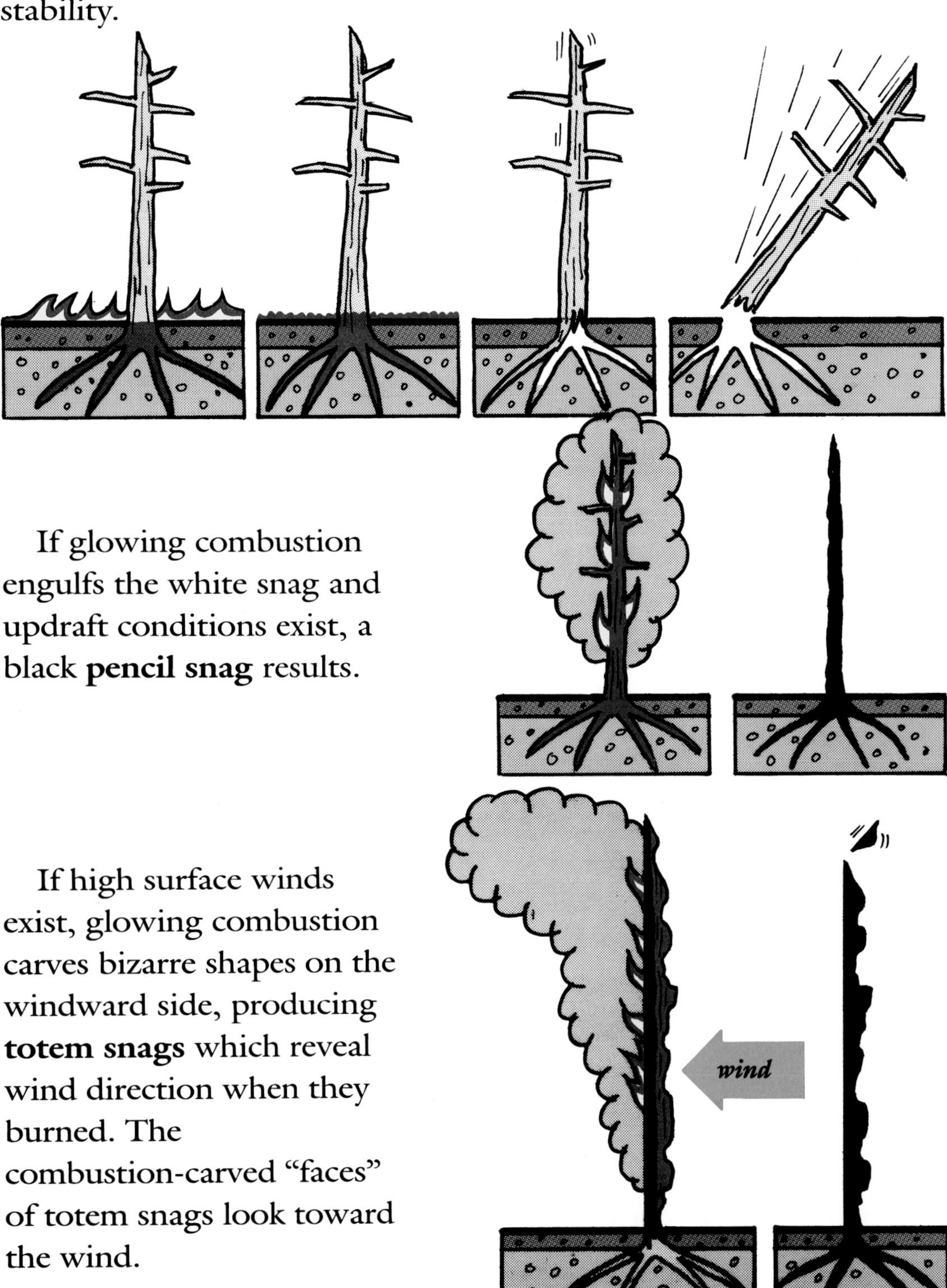

If glowing combustion engulfs the white snag and updraft conditions exist, a black **pencil snag** results.

If high surface winds exist, glowing combustion carves bizarre shapes on the windward side, producing **totem snags** which reveal wind direction when they burned. The combustion-carved "faces" of totem snags look toward the wind.

WIND EFFECT ON BURNING OF DEAD AND LIVING TREES

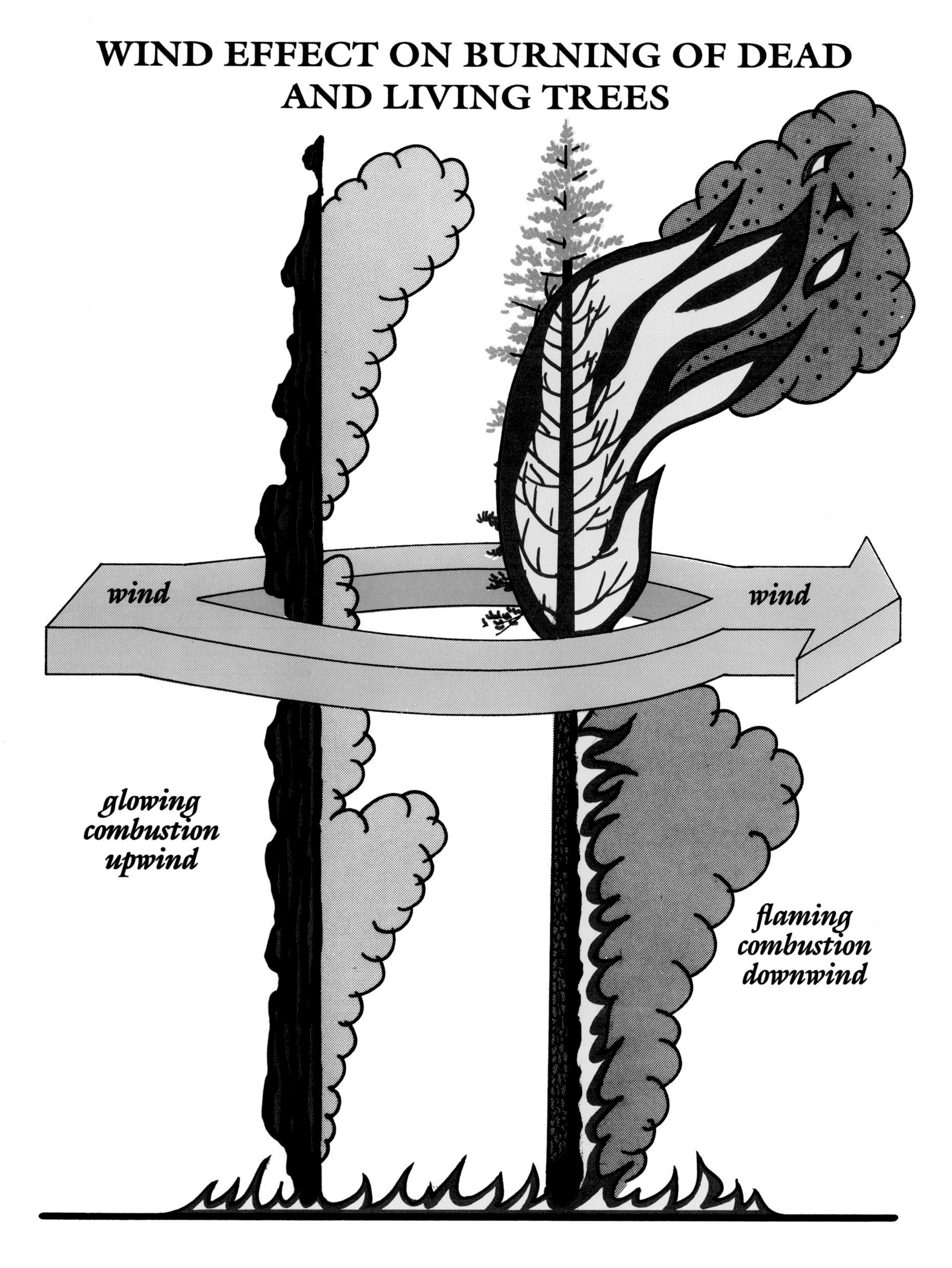

Surface wind intensifies glowing combustion on the upwind side of snags, producing **totems**. Conversely, it intensifies flaming combustion on the downwind side of living trees, producing **two-tones.**

CONVECTION COLUMNS AND LARGE FIRES

Regional weather determines the behavior of *small* fires. The heat of *large* fires creates its own local weather by the heat transfer mechanism of **convection**.

The parcel of air immediately above a fire contains highly energized molecules of hot gas. The highly energized molecules push outward and upward into surrounding cooler air, forming a hot, lightweight "bubble" of air. The surrounding ocean of cooler air flows in beneath the hot "bubble" to lift or "buoy" it upward into the sky.

The bubbles connect to form a continuous column of rising hot gases within a chimney of cooler air. The hot molecules rise until they cool to a temperature equal to surrounding air, which may be 5 to 7 miles above sea level.

The indrafts dramatically increase the rate of burn, and the updrafts carry smoke and ash high into the atmosphere, where they disperse widely and alter regional weather.

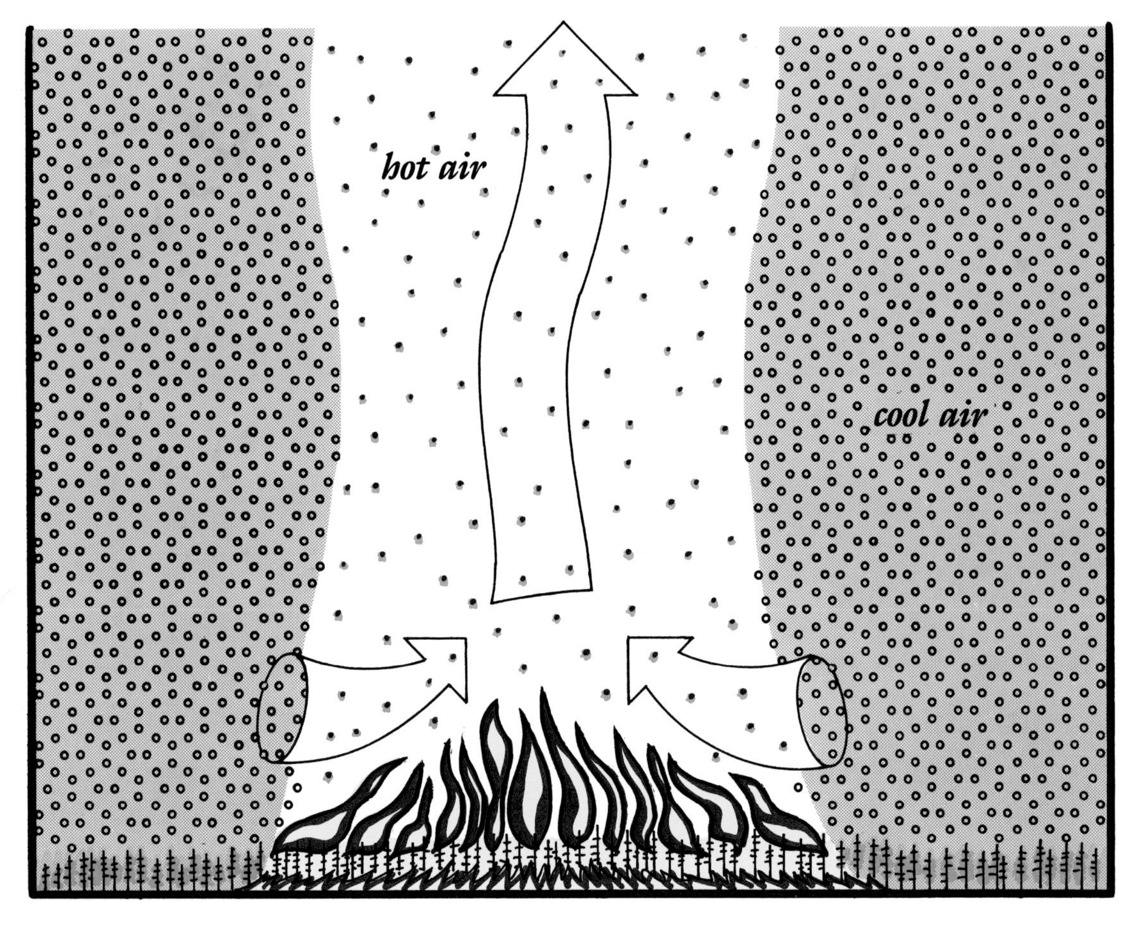

With surprising violence, plants demonstrate how much sunshine they have captured with their **producer reaction** when they suddenly release that energy as a **fire storm**, nature's most spectacular **consumer reaction**.

The transition from small to large fire is heralded by obvious indrafts, directional change of flames, increased burning rates (500%), and intense radiation. The fully developed convection column over a large fire generates the **fire storm**. Holocaust conditions exist within these storms.

A fat cumuliform cloud caps the column when the hot air carrying combustion by-products (CO_2 and H_2O) condenses at high altitudes.

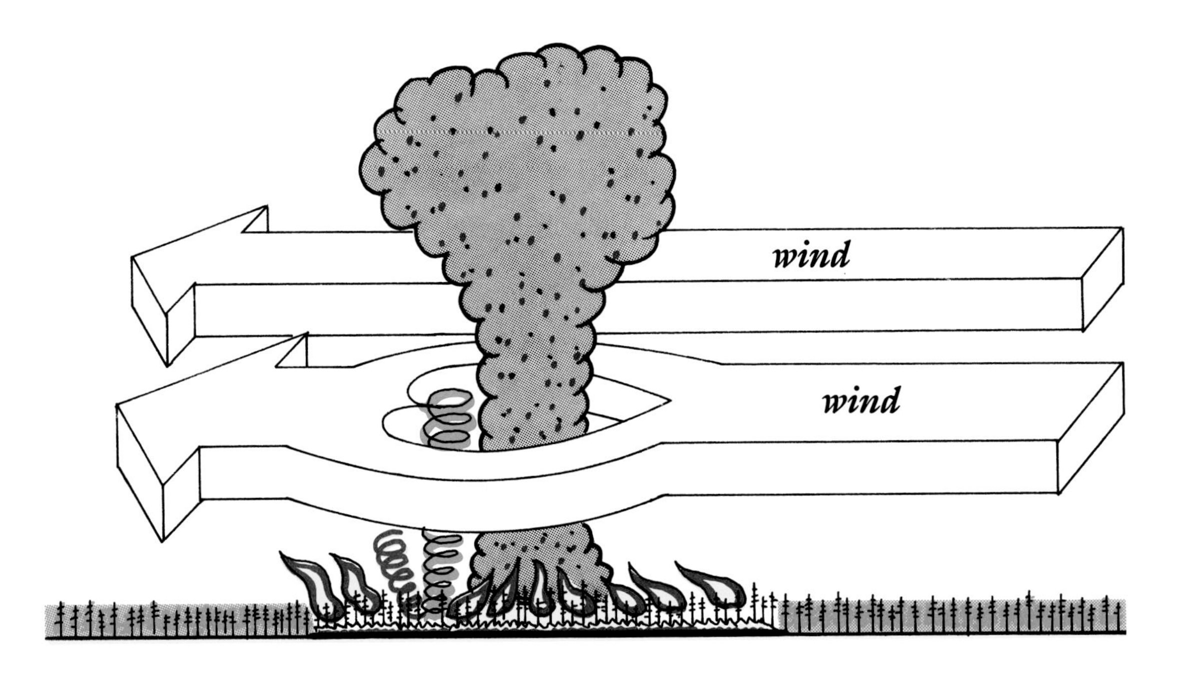

Like a boulder in a stream, strong convective columns behave as pillars and force the surface winds to diverge about them. The column produces turbulent eddies and whirlpools of air on its downwind side, resulting in dangerous and unpredictable fire behavior.

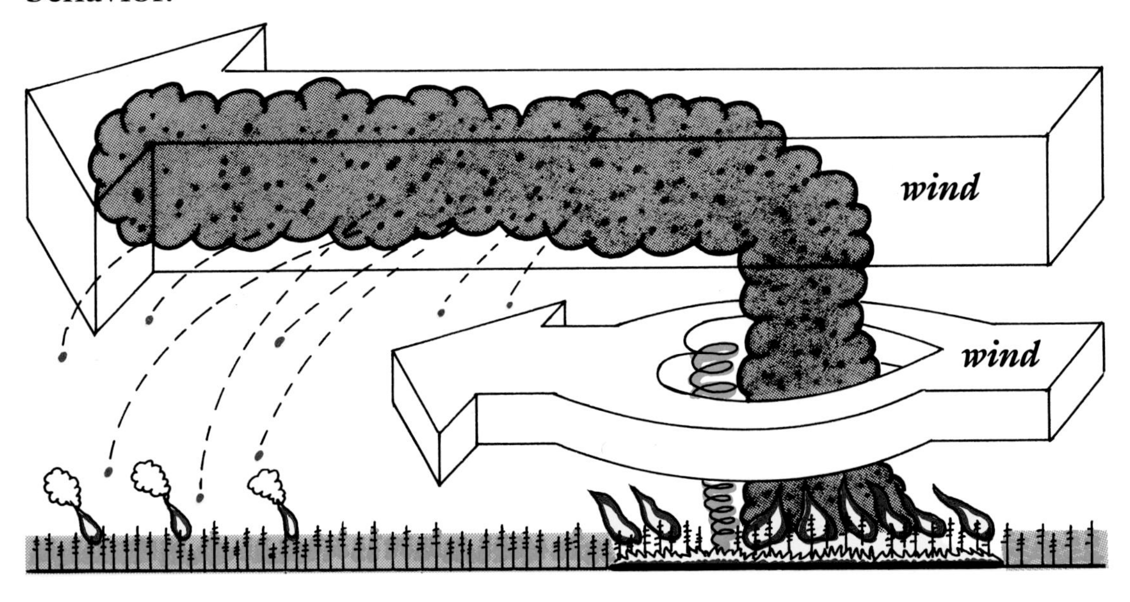

When winds are strong enough to break or fracture the column, it portends very severe spot fire activity and rapid fire spreading. This fire type may eject a rain of firebrands which land at a rate of thousands per acre.

WHIRLWINDS OF FIRE

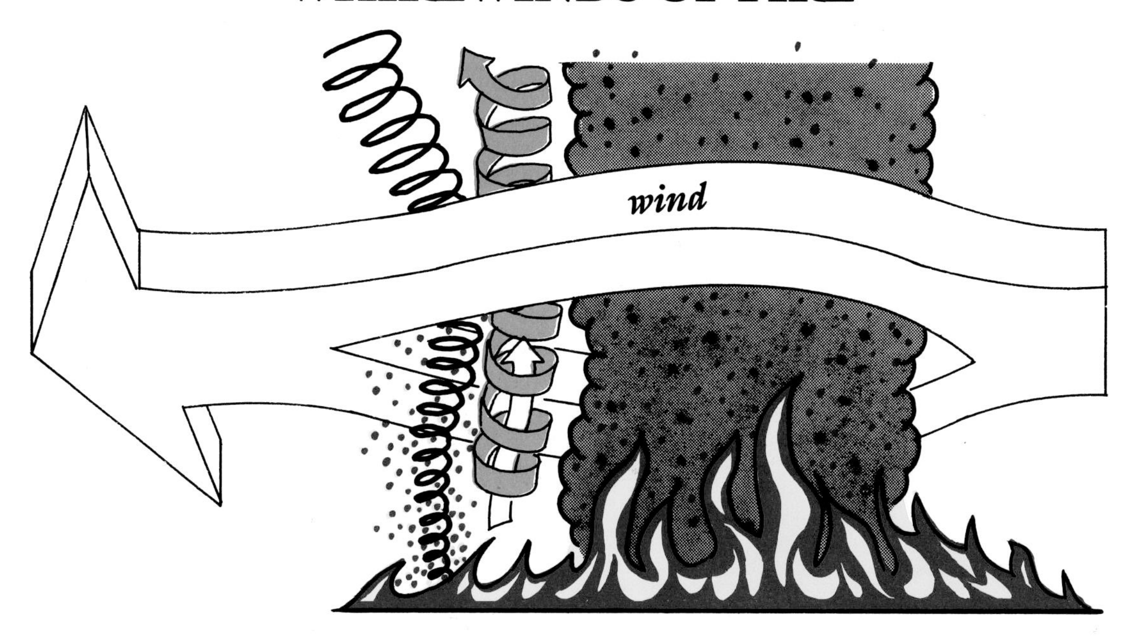

Fires generate convective updrafts, as previously discussed. If some force gives the updrafts a spin, a whirlwind develops. Large fires may produce tornado-like whirlwinds that seem to originate in the cloud on the turbulent downwind side of the convective column. These are rare, but may be very destructive, and did much of the damage in the famous Chicago fire of 1871.

More common are the dustdevil-sized fire whirls that develop from the convection column over intensely burning fuel. Like a twirling ice skater who tucks in her arms to spin faster, the rising column of air suddenly shrinks in diameter and starts rotating much faster. Burn rates increase 500%, which rapidly depletes the available fuel and abbreviates the life span of the fire whirl.

SMOKE

Smoke emitted by flaming fuel consists of two types: **gray pyrolysis smoke** from flame-radiated pre-ignition of unburned wood, and **black sooty smoke** from inside the broken flame bag of burning fuel *(see pages 22–23)*.

When the flames die, heat from glowing combustion preheats the unburned wood and produces **pyrolysis smoke**. A flameless fire of glowing combustion emits 800% more smoke than a flaming fire.

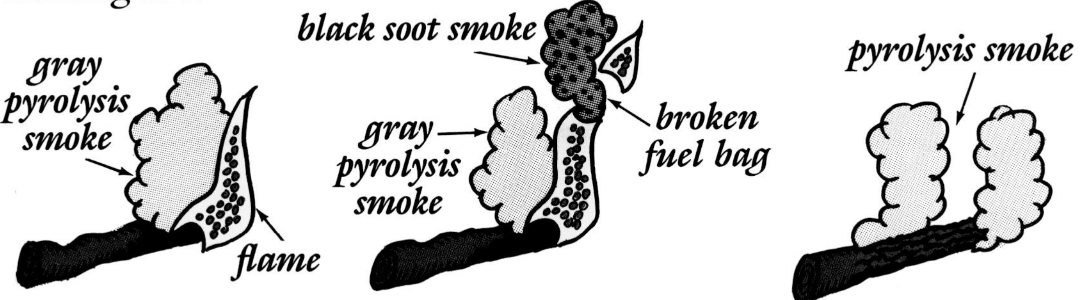

Black sooty smoke frequently accompanies the turbulent behavior of large fires. Powerful and turbulent updrafts rip off bits of flame, which lift or buoy skyward. Immense volumes of sooty smoke pour from the openings in the torn bag of burning fuel until its walls reform. The rising "bubbles" of golden flame cool rapidly, and their contained soot turns from orange to black in a flash.

In most wildland forest fires the smoke consists of 50% tar droplets of pyrolysis and 50% soot and ash particles. Wildland fire produces two or three tons of smoke particles per acre if the fuel loading is heavy.

IV. AFTERMATH AND CAMPFIRE STORY

"The living arises from the dead, the dead from the living."
—Plato

Wandering about in the ashes of a large fire enlightens the inquisitive mind and piques the imagination, like being miniaturized to explore the inside of a campfire. The embers and ashes reveal the story of all phases of combustion. They are the sole survivors of an immense chemical reaction that converted millions of tons of fuel to gases of carbon dioxide and water. Like magic, solid wood changes to gas and floats invisibly away.

FIRE CLASSIFICATION SYSTEM

After wildland fires, forest caretakers analyze the fire's aftermath in order to predict the effects on wildlife, streams, and future foliage. They first examine the amount of foliage involved in the fire, and then carefully evaluate the degree of soil heating and its effect on foliage regeneration.

The type of forest and its regeneration time depend on the degree of soil heating. The depth, intensity, and duration of soil heating determines the survival of seeds and organisms that ultimately replace the forestscape.

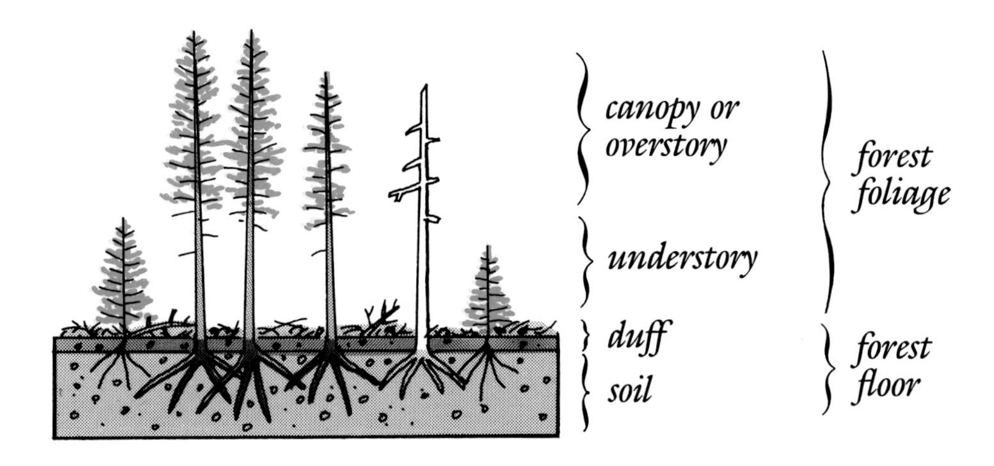

A very simple classification system based on visual clues evolves from much more intricate systems (see *Glossary: Forest floor burn*). Casual visitors can easily use this system.

FIRE CLASSIFICATION

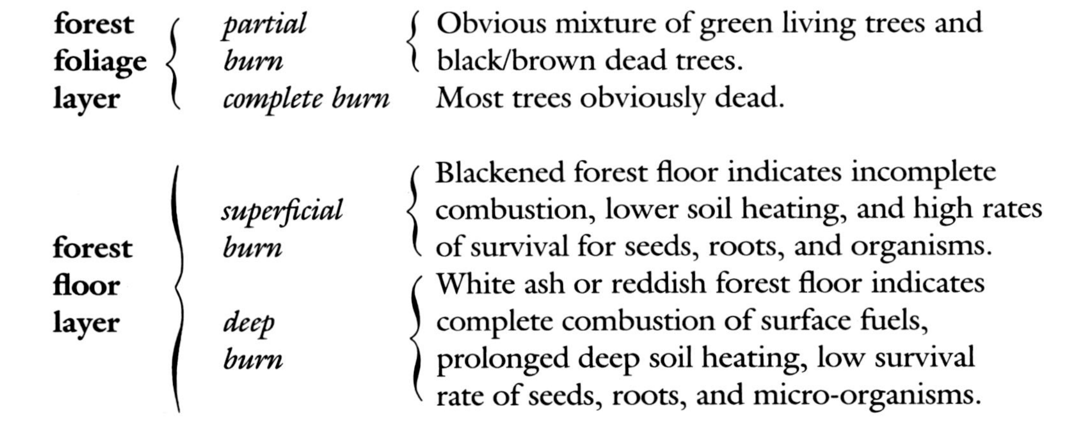

forest foliage layer	*partial burn*	Obvious mixture of green living trees and black/brown dead trees.
	complete burn	Most trees obviously dead.
forest floor layer	*superficial burn*	Blackened forest floor indicates incomplete combustion, lower soil heating, and high rates of survival for seeds, roots, and organisms.
	deep burn	White ash or reddish forest floor indicates complete combustion of surface fuels, prolonged deep soil heating, low survival rate of seeds, roots, and micro-organisms.

Flaming combustion of aerial and surface fuels, though dramatic in appearance, only minimally heats the soil layers of forest floor, because much of the searing energy of the reaction zone is directed away from the forest floor.

Glowing combustion of surface and ground fuels focuses much more of its reaction zone heat toward the soils, and for a very long time. This prolonged and directed heat impulse deeply heats the soils and kills many more life forms.

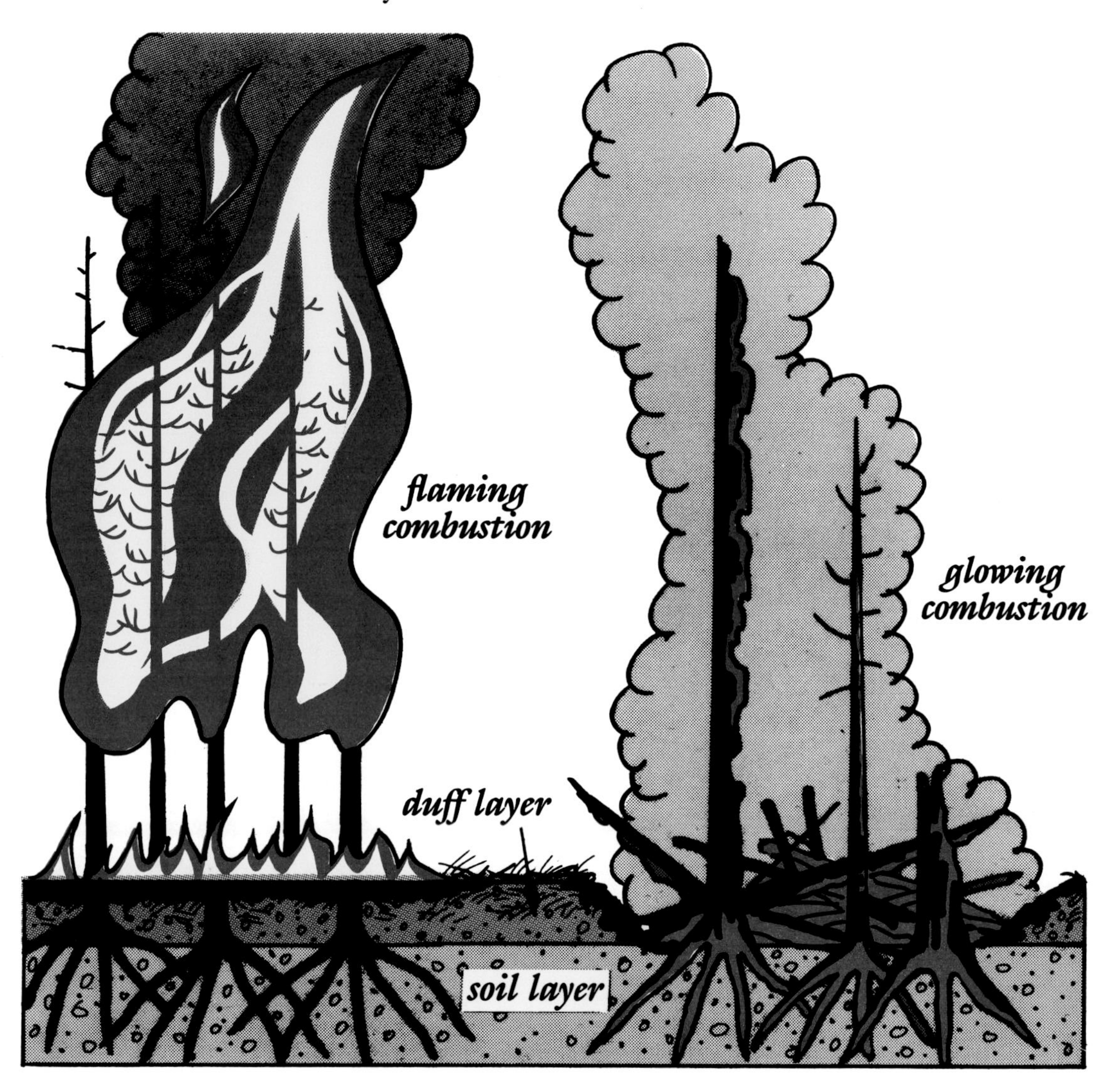

FLAMING COMBUSTION

Superficial burn, black forest floor of incomplete combustion.

GLOWING COMBUSTION

Deep burn, whitish forest floor of more complete combustion.

MOSAIC BURN

Viewed from a distance, forest foliage burns in a patchwork pattern called a **mosaic**. This irregular burning process contributes immensely to biological diversity.

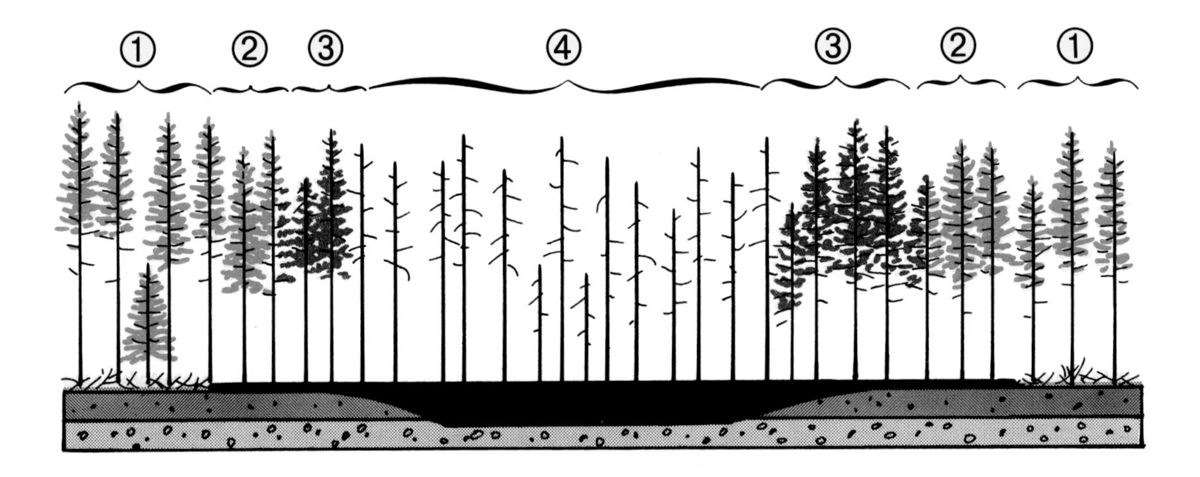

A cross-section through a typical mosaic area shows areas of differing burn types.

① Forest foliage: **unburned**. Forest floor: **unburned**.
② Forest foliage: **unburned**. Forest floor: **superficial burn**.
③ Forest foliage: **partial burn**. Forest floor: **superficial burn**.
④ Forest foliage: **complete burn**. Forest floor: **superficial and occasional deep burn**.

WHITE LINES AND CROSSES

Previously fallen dead trees may burn up completely because they contain little water. Long after the flames pass, they continue to burn by glowing combustion. They burn from underside to topside because ash easily falls away from the combustion reaction zone on the bottom side, permitting ongoing combustion.

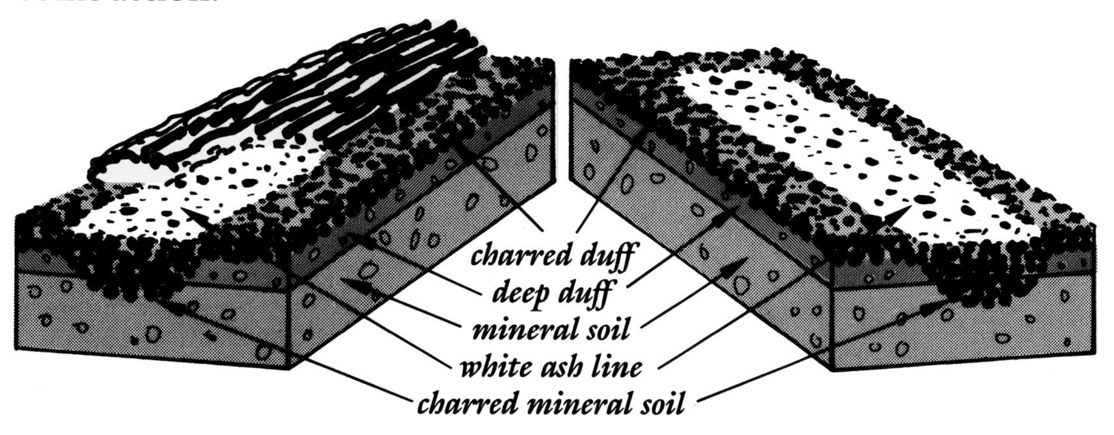

WHITE LINE

The downward focused and prolonged pulse of heat associated with glowing combustion of downed heavy fuels produces deep heating of the soil and may kill most life forms within its focus. Complete combustion leaves only mineral ash lines.

WHITE CROSS

flaming combustion

glowing combustion

white ash lines of complete combustion

MINI-MOSAIC

If a grove of trees burns intensely in a convective manner, updrafts and indrafts determine the intensity of burn and the direction of flames. During the strong convective stage of flaming combustion, indrafting air molecules direct the flames toward the center of the burning grove and create a circular burning pattern.

Interior trees burn and char entirely. Perimeter trees burn partially, and frequently in a "two-tone" producing manner.

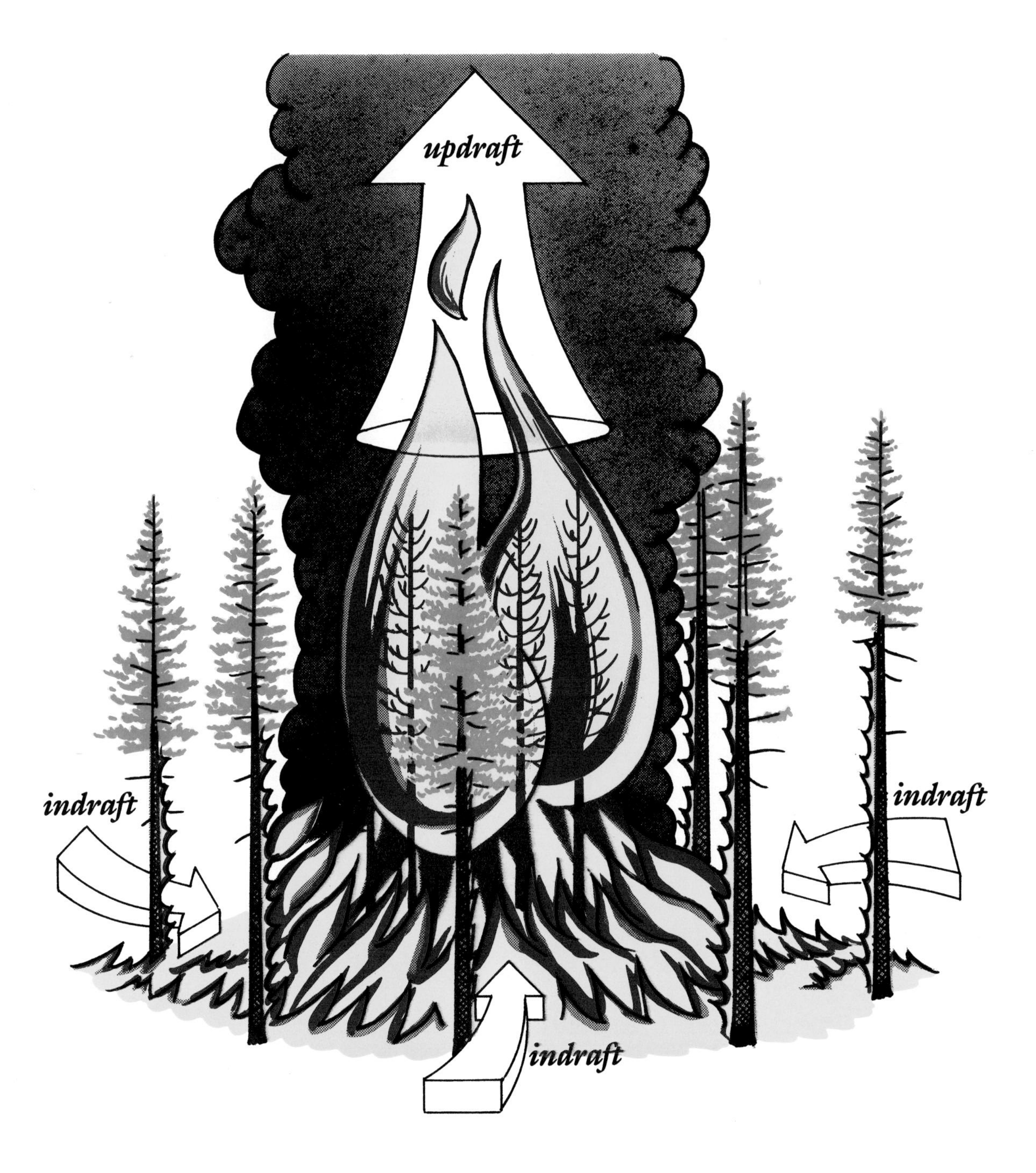

After the fire a micro-mosaic pattern remains, consisting of partially charred two-tone trees encircling a cluster of completely charred and burned trees. These micro-mosaic burns reveal the direction of convective updrafting and indrafting of wind by examination of the tree trunks.

FIRESCAPES

① This mature forest scene contains unburned trees, snags, heavy ground fuels, and small young trees.

② This firescape displays partially burned trees with dead and scorched needles. Charred litter suggests a superficial forest floor burn. White ash lines indicate complete combustion of heavy surface fuel and prolonged deep heating of soil. **Watch out for dangerous gray snags with burned-out root systems.**

③ This firescape reveals the passage of crown and surface fires. This thoroughly charred and burned forest follows intense flaming combustion of fine fuels. Black surface of burned litter suggests superficial soil heating. Pencil snags developed in windless or updrafting conditions. Numerous white lines and crosses tell of prolonged glowing combustion of heavy surface fuels and deep soil heating. Numerous pencil snags with burned-out root systems. **These snags can kill people!**

④ A white ash firescape indicates fearfully intense and complete combustion of aerial and surface fuels. The white remains of incinerated duff and heavy fuels cover a scorched mineral soil, indicating deep forest floor burn with high soil heating.

⑤ High surface winds sculpted this mixed firescape of partially burned two-tone trees and "villages" of exceedingly dangerous totem snags, whose tops snap off in the wind. Totems and two-tones warn of danger!

WAS THE TREE ALIVE OR DEAD WHEN IT BURNED?

Dead, dry, heavy fuel burns by glowing, leaving a **shiny black** and frequently **cobbled** surface.

Living, wet, heavy fuel burns by flaming, leaving a **dull black** surface of charred bark.

dead tree

glowing combustion: shiny, black, cobbled

live tree

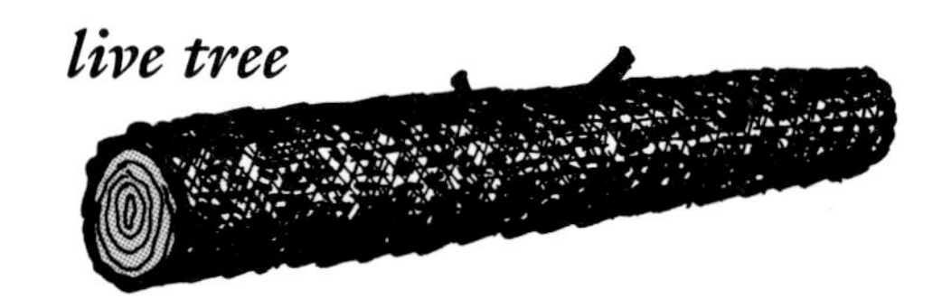

flaming combustion: dull black surface

CAMPFIRE STORY

A teepee of dried pine wood covers a loose pile of equally dry kindling of pine needles and twigs.

The pilot heat preheats the kindling, which signals **pre-ignition phase** by its production of tarry smoke. The fuel cloud forms.

The pilot heat ignites the fuel cloud, which bursts into flames of the **combustion phase**. Kindling flames in turn preheat the larger fuel to pre-ignition. If all goes well, the pilot heat of the flaming kindling ignites the new fuel cloud formed by pine wood.

The multi-shaped fuel clouds ignite as the kindling dwindles to glowing combustion and ash formation.

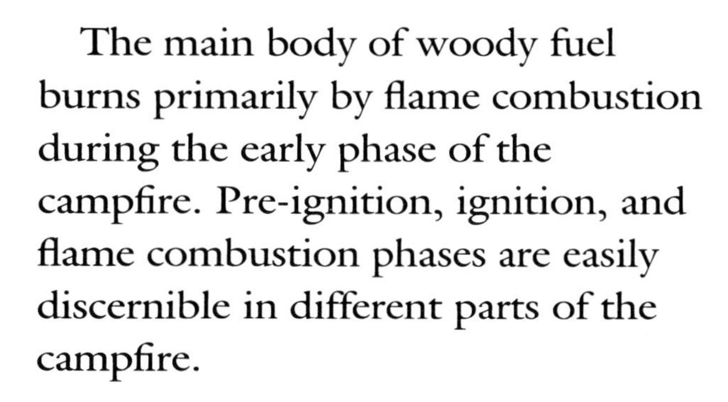

The main body of woody fuel burns primarily by flame combustion during the early phase of the campfire. Pre-ignition, ignition, and flame combustion phases are easily discernible in different parts of the campfire.

The later campfire has exhausted most of its gas fuels, and much of the wood cellulose has pyrolyzed to smoke and charcoal, which glowing combustion efficiently converts to carbon dioxide and water.

Cellulose pyrolysis deep inside the larger fuel disrupts the long cellulose molecules that provide strength and integrity to the wood. The heat-weakened cellulose permits the wood to collapse into itself.

Gray mineral ash coats the top surface, and it must be constantly knocked off to prevent smothering the coals.

The fire goes out, leaving remains of unburned charcoal bits, a few wood butts, and a fluffy pile of mineral ash. Most of the cellulose fuel has reacted with oxygen to form carbon dioxide and water. It's time for bed.

FIRE GLOSSARY

Many words in this book derive their definitions from the visual impressions implied from the illustrations. This glossary gives more precise definition to those assumptions.

Aerial fuels: The standing and supported forest combustibles not in direct contact with the ground, and consisting mainly of foliage, twigs, branches, stems, and bark.

Backing fire: A wildfire burning into or against the wind, or down the slope without the aid of wind.

Blow-up: A sudden increase in fire intensity and rate of spread, often associated with violent convection.

Classification systems for forest fires:

Forest foliage burn—partial or complete.

Forest floor burn—

Light burn: A degree of burn which leaves the soil covered with partially charred organic material; large fuels are not deeply charred. (Comparable to human first-degree burn).

Moderate burn: Degree of burn in which all organic material is burned away from the surface of the soil layer, which is not discolored by heat. Any remaining fuel is deeply charred. Organic matter remains in the soil immediately below the surface. (Comparable to human second-degree burn.)

Severe burn: Degree of burn in which all organic material is burned from the soil surface, which is discolored by heat, usually to red. Organic material below the surface is consumed or charred. (Comparable to human third-degree burn.)

Classification system for soil heating:

Low: Heating of mineral soil insignificant.

Moderate: Up to 5 cm of mineral soil charred.

High: Greater than 10 cm of mineral soil charred and surface often reddish due to mineral oxidation.

Severe: Soil crystallized to a hardened surface of larger mineral crystals.

Combustion: Consumption of fuels by oxidation, evolving heat and generally flame (neither necessarily sensible) and/or incandescence.

Conduction: Transfer of heat from molecule to molecule of a solid substance from region of higher temperature to region of lower temperature.

Conflagration: A raging, destructive fire. Often used to connote such a fire with a moving front, as distinguished from a fire storm.

Crown: The upper part of a tree or other woody plant, carrying the main branch system and foliage.

Crown fire: A fire that advances from top to top of trees or shrubs, more or less independently of the surface fire. Sometimes crown fires are classed as either running or dependent, to distinguish the degree of independence from the surface fire.

Dead fuels: Fuels having no living tissue in which the moisture content is governed almost entirely by atmospheric moisture (relative humidity and precipitation), air temperature, and solar radiation.

Decomposition: The physical breakdown of complex materials (organic or inorganic) into constituent parts.

Duff: Forest floor material composed of litter in various stages of decomposition. E.g. **litter** or **(L)** layer, **decomposing litter** or **(F)** fermentation layer, **organic soil** or **(H)**, humus layer.

Fine fuels/flash fuels: Fuels such as grass, leaves, draped pine needles, ferns, tree moss, and some slash that ignites readily and burns rapidly.

Firebrand: Any burning material, such as leaves, wood, and glowing charcoal or sparks, that could start a forest fire.

Fire storm: Violent convection caused by a large continuous area of intense fire; often characterized by destructively violent, surface indrafts, a towering convection column, long-distance spotting, and sometimes by tornado-like vortices.

Fire triangle: An instructional aid in which the sides of a triangle are used to represent the three factors (oxygen, heat, and fuel) necessary for combustion and flame production. When any one of these factors is removed, flame production ceases.

Fire whirl: A spinning vortex column of ascending hot air and gases rising from a fire and carrying aloft smoke, debris, and flame. Fire whirls range from a foot or two in diameter to small tornados in size and intensity. They may involve the entire area or only a hot spot within the fire area.

Flame: A mass of gas undergoing rapid combustion, generally accompanied by the evolution of sensible heat and incandescence.

Flaming combustion: Luminous oxidation of the gases evolved from the thermal decomposition of the fuel.

Flaming front: That zone of a moving fire where the combustion is primarily flaming. Behind this flaming zone combustion is primarily glowing.

Flanking fire: That part of a fire that is roughly parallel to the main direction of the heading and backing fires.

Forest: (1) Generally, an ecosystem characterized by a more or less dense and extensive tree cover. (2) More particularly, a plant community predominantly of trees and other woody vegetation growing more or less closely together, and composed of an overstory (canopy) and understory.

Forest floor: The duff and soil layers supporting the forest.

Fuel: Combustible material.

Glowing combustion: Oxidation of a solid surface accompanied by incandescence, sometimes evolving small flame above it (e.g. blue flame of CO oxidation).
Ground fire: Fire that burns organic material in the subsurface duff and mineral soil layers.
Ground fuel: All combustible materials below the surface litter, including duff, tree or shrub roots, punky wood, peat, and sawdust, that normally support glowing combustion without flame (smoldering combustion).
Heading fire: A fire spreading with the wind or uphill.
Heat transfer: The process by which heat energy is imparted from one body to another through conduction, convection, or radiation.
Heavy fuels: Fuels of large diameter, such as snags, logs, and large branch-wood, or of a peaty nature, that ignite and burn more slowly than flash fuels.
Ladder fuels: Fuels which provide vertical continuity between strata. Fire is able to carry from surface fuels into crowns with relative ease.
Large fire: A fire burning with a size and intensity such that its behavior is determined by interactions between its own convection column and weather conditions above the surface.
Light burn: see **classification systems**.
Litter: The top layer of the forest floor, composed of loose debris of dead sticks, branches, twigs, and recently fallen leaves or needles, little altered in structure by decomposition.
Living fuels: Naturally occurring fuels in which the moisture content is physiologically controlled within the living plant.
Mineral soil: Soil layers below the predominantly organic soils of the duff layer, which contain few combustibles.
Moderate burn: see **classification systems**.
Parts of a fire: On typical free-burning fires the spread is uneven, with the main spread moving with the wind or upslope. The most rapidly moving portion is designated the head of the fire, the adjoining portions of the perimeter at right angles to the head are known as the flanks, and the slowest moving portion is known as the rear, or back.
Pyrolysis: The thermal or chemical decomposition of fuel at an elevated temperature.
Scorch: Permanent discoloration caused by heating, which may be black if charred, or yellowish-brown if killed by heating but not charred or burned.
Severe burn: see **classification systems**.
Smoldering combustion: Flameless combustion in a fuel-rich and very oxygen-poor environment within stumps, dead logs, underground roots, or other tightly packed fuels.

Snag: A standing dead tree or standing portion from which at least the leaves or needles and smaller branches have fallen. Often called a stub if less than 6 m tall.

Spot fire: Fire set outside the perimeter of the main fire by flying sparks or embers.

Surface fire: Fire that burns only surface litter, other loose debris of the forest floor, and small vegetation.

Surface fuel: The loose surface litter on the forest floor, normally consisting of fallen leaves or needles, twigs, bark, cones, and small branches that have not yet decayed sufficiently to lose their identity. Also grasses, shrubs, and tree reproduction less than 1 m in height, heavier branchwood, down logs, stumps, seedlings, and forbs (weeds) interspersed with or partially replacing the litter.

Torching: A tree (or small clump of trees) is said to "torch" when its foliage ignites and flares up, usually from bottom to top.

Volatiles: Readily vaporized organic materials which, when mixed with oxygen, are easily ignited.

Wildland fire: Any fire occurring on wildland except a fire under prescription.

Wildland: An area in which development is essentially nonexistent, except for roads, railroads, powerlines, and similar transportation facilities. Structures, if any, are widely scattered, and are primarily for recreation purposes.

RECOMMENDED READING

Most references require the reader to possess a familiarity with the sciences of physics, chemistry, and biology.

Simplified information on the combustion process and the structure and function of flame, as presented in this book, is not available from non-technical resources, and was obtained by the author through personal communication with appropriate research scientists.

TEXTBOOKS

Chandler, C., P. Cheney, P. Thomas, L. Trabaud, and D. Williams. *Fire in Forestry, Vols. I, II.* NY: John Wiley and Sons, 1983.

Davis, K. P. *Forest Fire: Control and Use.* NY: McGraw-Hill, 1959.

Pyne, S. J. *Introduction to Wildland Fire: Fire Management in the United States.* Princeton: Princeton University Press, 1984.

Wright, H. A., and A. W. Bailey. *Fire Ecology: The United States and Southern Canada.* NY: John Wiley and Sons, 1982.

Fristrom, R. M., and A. A. Westenberg. *Flame Structure.* Chapman and Hall, 1953.

TECHNICAL PUBLICATIONS

Proceedings—Symposium and Workshop on Wilderness Fire, General Technical Report INT-182, Intermountain Forest and Range Experiment Station, Ogden, UT 84401, 1985.

Protecting People and Homes from Wildfire in the Interior West: Proceedings of Symposium and Workshop, General Technical Report 251, Intermountain Research Station, Ogden, UT 84401, 1988.

SPIRITUAL PUBLICATIONS

McHugh, J., and L. Harris. *Journey to the Moon.* Millbrae, CA: Celestial Arts, 1974.

ABOUT THE AUTHOR

William H. Cottrell Jr. spent his childhood in the midwest, a bit east of Kansas City, Missouri. Colorado State University granted him a B.S. in Zoology prior to his attending medical school at the University of Missouri. Military experience in Alaska preceded an orthopaedic surgical residency at the University of Southern California. He chose the Sierra foothill community of Placerville in which to live with his family.

The private practice of surgery controlled much of his time until a disabling rock-climbing accident provided him with time to write. The ability to present complex scientific information by combining accurate graphics and simplified text developed from the patient-surgeon relationship. He specializes in providing readers with an accurate impression of difficult subject matter, regardless of their levels of education or familiarity with the subject.

His successful first book *Born of Fire*, described the volcanic origins of Yellowstone Park. *The Book of Fire* is his second book.

Design, graphics, and composition by
Comstock Bonanza Press and Dwan Typography.
The type family is Galliard, digitally expanded on the Linotron 202.